TRANSPORTATION INFRASTRUCTURE - ROADS, HIGHWAYS, BRIDGES, AIRPORTS AND MASS TRANSIT

SURFACE TRANSPORTATION

MAP-21 PROVISIONS AND TIFIA CREDIT ASSISTANCE

Transportation Infrastructure - Roads, Highways, Bridges, Airports and Mass Transit

Additional books in this series can be found on Nova's website under the Series tab.

Additional E-books in this series can be found on Nova's website under the E-book tab.

Transportation Issues, Policies and R&D

Additional books in this series can be found on Nova's website under the Series tab.

Additional E-books in this series can be found on Nova's website under the E-book tab.

TRANSPORTATION INFRASTRUCTURE - ROADS, HIGHWAYS, BRIDGES, AIRPORTS AND MASS TRANSIT

SURFACE TRANSPORTATION

MAP-21 PROVISIONS AND TIFIA CREDIT ASSISTANCE

DONNA NEIL
EDITOR

nova publishers
New York

For permission to use material from this book please contact us:
Telephone 631-231-7269; Fax 631-231-8175
Web Site: http://www.novapublishers.com

Library of Congress Cataloging-in-Publication Data

ISBN: 978-1-62417-431-5

Published by Nova Science Publishers, Inc. †New York

Contents

PREFACE

On July 6, 2012, President Barack Obama signed the Moving Ahead for Progress in the 21st Century Act (MAP-21; P.L. 112-141). The act authorized spending on federal highway and public transportation programs, surface transportation safety and research, and some rail programs and activities through September 30, 2014. MAP-21 authorized roughly \$105 billion for FY2013 and FY2014 combined. This book examines the MAP-21 Act and the Transportation Infrastructure Finance and Innovation Act (TIFIA) program, which is designed to fill market gaps and leverage substantial nonfederal investment by providing federal credit assistance to help finance surface transportation projects including highway, transit, rail, and intermodal projects.

Chapter 1 – On July 6, 2012, President Barack Obama signed the Moving Ahead for Progress in the 21st Century Act (MAP-21; P.L. 112-141). The act authorized spending on federal highway and public transportation programs, surface transportation safety and research, and some rail programs and activities through September 30, 2014. MAP-21 authorized roughly \$105 billion for FY2013 and FY2014 combined. It also extended FY2012 surface transportation authorizations to the end of the fiscal year, raising the total authorization to approximately \$118 billion.

Most of the funding for surface transportation bills has been drawn from the highway trust fund (HTF) since its creation in 1956, but the HTF, which receives revenue mainly from federal motor fuel taxes, has experienced declining revenue due to a sluggish economy and improvements in vehicle fuel efficiency. For the past several years, HTF revenue has been insufficient to finance the government's surface transportation programs, leading Congress to delay reauthorization for 33 months following expiration of the last multi-year reauthorization. Although Congress was unable to agree on a long-term solution to the HTF revenue issue, MAP-21 did provide for the transfer of sufficient general fund revenues to the HTF to fund a two-year bill.

MAP-21 made major changes in the programmatic structure for both highways and public transportation and included initiatives intended to increase program efficiency through performance-based planning and the streamlining of project development. Among its major provisions, MAP-21 included:

- for the federal-aid highway program, research, and education, authorizations for FY2013 of $40.96 billion and for FY2014 of $41.03 billion;
- for public transportation, authorizations for FY2013 of $10.58 billion and for FY2014 of $10.7 billion;
- for the Transportation Infrastructure Financing and Innovation Act (TIFIA), which provides credit assistance for surface transportation projects, a significant expansion that could provide credit support of up to $690 million for FY2013 and $9.2 billion for FY2014;
- major program restructuring, which reduced the number of highway programs by two-thirds and consolidated public transportation programs as well;
- more distribution of funding via apportionment to the states and less discretionary funding via the Department of Transportation (DOT) to individual projects;
- no project earmarks;
- no equity program, instead basing the distribution of highway funding on the FY2012 distribution such that each state will likely receive as much federal highway funding as its highway users paid to the highway account of the HTF; and
- changes in the National Environmental Policy Act (NEPA) compliance process intended to accelerate project delivery.

Chapter 2 – Created in 1998, the TIFIA program is designed to fill market gaps and leverage substantial nonfederal investment by providing federal credit assistance to help finance surface transportation projects including highway, transit, rail, and intermodal projects. Since 2008, demand for the program has surged, annually exceeding budget resources for the program by a factor of more than 10 to 1. Given the increased demand and recent proposals to expand and modify the program, GAO was asked to review (1) the characteristics of TIFIA projects and how DOT tracks progress toward the program's goals, (2) the process DOT used to evaluate and select projects that submitted LOIs to apply for credit assistance in fiscal years 2010 and 2011, (3) the factors that affect project sponsors' decisions about whether to seek TIFIA credit assistance, and (4) the options proposed to modify the program. GAO reviewed laws and program guidance; interviewed DOT officials, project sponsors, and advisors involved in procuring credit assistance; and surveyed all state departments of transportation and other recent applicants about the TIFIA program.

In: Surface Transportation
Editor: Donna Neil
ISBN: 978-1-62417-431-5

Chapter 1

SURFACE TRANSPORTATION FUNDING AND PROGRAMS UNDER MAP-21: MOVING AHEAD FOR PROGRESS IN THE 21ST CENTURY ACT (P.L. 112-141)*

Robert S. Kirk, John Frittelli, Linda Luther, William J. Mallett and David Randall Peterman

SUMMARY

On July 6, 2012, President Barack Obama signed the Moving Ahead for Progress in the 21st Century Act (MAP-21; P.L. 112-141). The act authorized spending on federal highway and public transportation programs, surface transportation safety and research, and some rail programs and activities through September 30, 2014. MAP-21 authorized roughly $105 billion for FY2013 and FY2014 combined. It also extended FY2012 surface transportation authorizations to the end of the fiscal year, raising the total authorization to approximately $118 billion.

Most of the funding for surface transportation bills has been drawn from the highway trust fund (HTF) since its creation in 1956, but the HTF, which receives revenue mainly from federal motor fuel taxes, has experienced declining revenue due to a sluggish economy and improvements in vehicle fuel efficiency. For the past several years, HTF

* This is an edited, reformatted and augmented version of Congressional Research Service, Publication No. R42762, dated September 27, 2012.

revenue has been insufficient to finance the government's surface transportation programs, leading Congress to delay reauthorization for 33 months following expiration of the last multi-year reauthorization. Although Congress was unable to agree on a long-term solution to the HTF revenue issue, MAP-21 did provide for the transfer of sufficient general fund revenues to the HTF to fund a two-year bill.

MAP-21 made major changes in the programmatic structure for both highways and public transportation and included initiatives intended to increase program efficiency through performance-based planning and the streamlining of project development. Among its major provisions, MAP-21 included:

- for the federal-aid highway program, research, and education, authorizations for FY2013 of $40.96 billion and for FY2014 of $41.03 billion;
- for public transportation, authorizations for FY2013 of $10.58 billion and for FY2014 of $10.7 billion;
- for the Transportation Infrastructure Financing and Innovation Act (TIFIA), which provides credit assistance for surface transportation projects, a significant expansion that could provide credit support of up to $690 million for FY2013 and $9.2 billion for FY2014;
- major program restructuring, which reduced the number of highway programs by two-thirds and consolidated public transportation programs as well;
- more distribution of funding via apportionment to the states and less discretionary funding via the Department of Transportation (DOT) to individual projects;
- no project earmarks;
- no equity program, instead basing the distribution of highway funding on the FY2012 distribution such that each state will likely receive as much federal highway funding as its highway users paid to the highway account of the HTF; and
- changes in the National Environmental Policy Act (NEPA) compliance process intended to accelerate project delivery.

MAP-21: Overview

Surface transportation authorization acts authorize spending on federal highway and public transportation programs, surface transportation safety and research, and some rail programs. The most recent multi-year authorization for federal surface transportation programs, the Moving Ahead for Progress in the

21st Century Act (MAP-21; P.L. 112-141), reauthorizes federal surface transportation programs and activities through September 30, 2014.

MAP-21 authorizes roughly $105 billion for FY2013 and FY2014 combined. It also extended FY2012 surface transportation authorizations to the end of the fiscal year, raising the act's total authorization to approximately $118 billion.

The comparatively short two-year authorization reflects the difficulty faced by Congress in funding surface transportation within a constrained budgetary environment. The highway trust fund (HTF) has provided most of the funding for surface transportation authorization bills since its creation in 1956. The HTF is supported mostly from taxes on gasoline and diesel fuel, but revenue has declined in recent years due to a sluggish economy and improvements in vehicle fuel efficiency. The taxes were last raised in 1993 and are fixed in terms of cents per gallon, so revenues do not rise along with motor fuel prices. Consequently, Congress faced a choice of cutting transportation spending, increasing motor fuels taxes, or providing general fund transfers to supplement the HTF revenues. In the end, Congress chose to transfer money from the general fund to the HTF to fund a two-year authorization.

Congress also made major changes in the structure and formula of the highway and transit programs. The number of highway programs was reduced by roughly two-thirds. Mass transit programs were also reduced through consolidation. Although the eligibilities of many former programs were absorbed into new or modified programs under MAP-21, the changes have the potential to increase efficiency. The six large core programs under the previous authorization, the Safe, Accountable, Flexible, Efficient Transportation Equity Act: A Legacy for Users (SAFETEA; P.L. 109-59), were consolidated into five. The new National Highway Performance Program absorbed the Interstate Maintenance, National Highway System, and Federal-Aid Highway Bridges programs. Off-system highway bridges were included in the Surface Transportation Program. The Congestion Mitigation and Air Quality Program (CMAQ) and the Highway Safety Improvement Program (HSIP) were both retained, and the Recreational Trails, Safe Routes to Schools, Transportation Enhancements, and certain other eligibilities were combined into a new core program, Transportation Alternatives (TA).

Over 90% of the highway program funding under MAP-21 is distributed by formula and is under the control of the state departments of transportation. The number of discretionary programs, formerly under the control of the Federal Highway Administration (FHWA), has been reduced. In addition,

MAP-21 is free of project earmarking. These changes continue a long-term shift away from detailed federal control of project decisions to a policy of limited federal oversight of project control exercised at the state and metropolitan levels.

Unlike previous authorizations, MAP-21 does not use separate formulas to distribute funds for individual programs. Instead, it applies a broad initial calculation across virtually all the formula programs and then assigns a percentage of the available funding to each formula program. The initial calculation sets each state's apportionment share under most of the formula programs. In place of the old formula factors, such as, for example, lane miles on principal arterial routes, MAP-21 relies on a performance-based approach to rewarding progress toward national goals.

MAP-21's use of the initial calculation eliminated the need for an equity bonus program to guarantee that each state would receive at least a proportion of its residents' payments into the HTF, as in SAFETEA.[1] Instead, MAP-21 sets each state's share in the initial distribution for FY2013 and FY2014 based on the apportionment it received in FY2012, adjusted in FY2014 to assure that each state receives 95% or more of its highway tax contribution to the HTF. This adjustment may not be necessary, as all states have received more in apportionments and allocations than their tax payments to the HTF in every year since FY2007.

MAP-21 also did not include the budgetary "firewalls" and spending guarantees that were maintained under both SAFETEA and its predecessor law, the Transportation Equity Act for the 21st Century (P.L. 105-178). These firewalls and spending guarantees made it difficult for surface transportation spending levels to be lowered during the annual appropriations process. Under MAP-21 the ability of the appropriators to lower authorized funding levels has been restored.

During the MAP-21 debate Congress considered funding sources other than grants. The act provides for a major expansion of the Transportation Infrastructure Financing and Innovation Act (TIFIA) and some modest changes in federal tolling provisions. TIFIA is touted as allowing for the leveraging of almost ten times the amount authorized. Tolling changes reaffirm the support for value and congestion pricing and allow for tolling of new Interstate route construction.

Public transportation is provided with slight spending increases under MAP-21. Transit programs have undergone significant structural change. The new State of Good Repair Program replaces the Fixed Guideway Modernization Program, and transit authorities are required to develop asset management systems. The Bus and Bus Facilities Program is funded at less than half the level of FY2012 and the funds are distributed by formula rather than being earmarked.

The federal program for freight transport proposed in the Senate surface transportation bill (S. 1813, 112th Congress) was not included in MAP-21 as enacted. The act did retain many planning provisions related to identifying infrastructure components most critical to freight, including the designation of the "primary freight network."

MAP-21 also seeks to accelerate transportation project delivery by altering the National Environmental Policy Act (NEPA) compliance process and other environmental reviews.[2] However, it is not certain that the enacted changes will speed project completion, and there is reason to think that some projects may face additional delay due to the new law.

Surface Transportation Finance and the Highway Trust Fund

The highway trust fund is financed from a number of sources including sales taxes on tires, trucks, buses, and trailers, as well as truck usage taxes. However, approximately 90% of trust fund revenue comes from excise taxes on motor fuels, 18.3 cents per gallon on gasoline and 24.3 cents per gallon on diesel. The HTF consists of two separate accounts—highway and mass transit. The highway account receives an allocation equivalent to 15.44 cents of the gasoline tax and the mass transit account receives the revenue generated by 2.86 cents of the tax.[3] Because the fuel taxes are set in terms of cents per gallon, rather than as a percentage of the sale price, their revenues do not increase with inflation. The fuel tax rates were last raised in 1993.

The period of sluggish economic performance that began in 2007 and improvements in vehicle fuel efficiency have reduced fuel tax revenue below the optimistic projections assumed in SAFETEA. The highway account has already required three transfers from the general fund totaling $29.7 billion,[4] without which the Federal Highway Administration (FHWA) might not have been able to pay states for work they completed. The mass transit account received a $4.8 billion general fund transfer in FY2010. Both accounts were projected to have insufficient resources to fulfill MAP-21 funding levels without further general fund transfers.

The finance provisions of MAP-21 met the need, providing general fund transfers of $6.2 billion and $12.6 billion for FY2013 and FY2014 respectively.[5] $2.4 billion of the accrued balance of the Leaking Underground

Storage Tank Fund was also transferred to the HTF for FY2012. According to Congressional Budget Office estimates, the highway account of the HTF will retain a prudent balance of $4.1 billion at the end of MAP-21 (i.e., FY2014). The mass transit account, however, is expected to have a balance of $0.5 billion, only half the amount considered prudent.[6]

Projections indicate that HTF revenues will continue to be inadequate to support baseline spending on surface transportation programs after FY2014. This suggests that a reauthorization of MAP-21 would require an increase in the existing fuel taxes, continuing expenditures from the general fund, or reductions in the scope of the federal surface transportation program.

Budget Control Act of 2011 (P.L. 112-25)

The Budget Control Act requires sequestration of certain funding authorizations in the event a special joint committee of Congress fails to reach an agreement on spending reductions. The Budget Control "Super Committee" announced in November 2011 that it had failed to reach such an agreement. However, exemptions to the sequester process under the Balanced Budget and Emergency Deficit Control Act of 1985, as amended (codified in 2 U.S.C. §905 (j)), likely mean that sequestration would not significantly reduce any surface transportation spending authorized for years beyond FY2012. The surface transportation programs and activities exempted, to the extent that their budgetary resources are subject to appropriations bill obligation limitations, are:

- federal-aid highways;
- highway traffic safety grants;
- National Highway Traffic Safety Administration (NHTSA) operations and research and National Driver Register;
- motor carrier safety operations, programs, and grants; and
- transit formula and bus grants.

The $739 million of annual contract authority that is typically exempt from the obligation limitation appears to be subject to sequestration. The Federal Transit Administration (FTA) New Starts program, supported with general fund revenues, also appears to be subject to sequestration.

HIGHWAYS

The Federal-Aid Highway Program (Highway Program) is an umbrella term for an array of programs administered by FHWA. The Highway Program is primarily state-run. The state departments of transportation (state DOTs) largely determine where and how money is spent, but have to comply with detailed federal planning guidelines. The state DOTs award the contracts and oversee project development and construction. States and/or local governments usually must provide a matching share, typically 10% for Interstate System projects and 20% for other roads.

Highway Program Terminology

Distribution of funds is FHWA notification of the availability of federal funds, usually for four years. The states do not actually receive federal money for highway project spending up front.

Apportionment is the distribution of funds among the states as prescribed by a statutory formula.

Allocation is an administrative distribution of funds (often for specific projects) under programs that do not have statutory distribution formulas.

Reimbursement occurs once a project is approved, the work is started, costs are incurred, and the state submits a voucher to FHWA. The reimbursable structure is designed to curb waste, fraud, and abuse.

Contract authority is a type of budget authority that is available for obligation even without an appropriation (although appropriators must eventually provide liquidating authority to pay the obligations).

Obligation of contract authority for a project by FHWA legally commits the federal government to reimburse the state for the federal share of a project. This can be done prior to an appropriation.[7]

Limitation on obligations, known as ObLim or Oblimit, is used to control annual FHWA spending in place of an appropriation. The ObLim sets a limit on the total amount of contract authority that can be obligated in a single fiscal year. For practical purposes, the ObLim is analogous to an appropriation.[8]

FEDERAL-AID HIGHWAYS

- MAP-21's highway and research titles authorize $81.99 billion over two years, $40.96 billion for FY2013 and $41.03 billion for FY2014 (see **Table 1**).
- MAP-21 eliminates the Equity Bonus Program. Instead the bill guarantees a state share based on SAFETEA and a 95 cent return (in FY2014) on each dollar that a state's highway users pay to the highway account of the HTF.
- MAP-21 does not include highway program earmarks.
- MAP-21 eliminates or consolidates most discretionary programs.

Table 1. Highway Authorizations: MAP-21 (contract authority from the highway account of the HTF, except as noted, in millions of dollars)

Program	FY2013	FY2014	Total
Title I: Federal-Aid Highways	37,477	37,798	75,275
Transportation Infrastructure Finance and Innovation	750	1,000	1,750
Program (TIFIA)			
Tribal Transportation Program	450	450	900
Federal Lands Transportation Program	300	300	600
Federal Lands Access Program	250	250	500
Territorial and Puerto Rico Highway Program	190	190	380
Federal Highway Administration Administrative Expenses	454	440	894
Emergency Relief	100	100	200
Projects of National & Regional Significance [Gen. Fund]	500	-	500
Construction of Ferry Boats and Ferry Terminal Facilities	67	67	134
Tribal High Priority Projects Program [Gen. Fund]	30	30	60
Total Authorizations: Division A	40,568	40,625	81,193
Federal-Aid Highway Program Obligation Limitation	39,699	40,256	79,955
Title II Research and Education	400	400	800
Total Authorizations	40,968	41,025	81,993

Source: Federal Highway Administration. For breakout of formula programs, see Table 2.

Notes: The $500 million authorized for Projects of National and Regional Significance for FY2013 and the annual $30 million for Tribal High Priority Projects can be expended only with an appropriation.

Formulas and Apportionments

Unlike SAFETEA and earlier authorization acts, MAP-21 does not use individual program formulas to determine each state's apportionments under each core program. Instead, MAP-21 has an annual authorization for each major programs and a single funding formula, as follows:

First, for FY2013, each state's "initial amount" is determined by multiplying the total amount available for apportionment for the year by each state's share of total nationwide apportionments and allocations received for FY2012. For FY2014, the total amount available for distribution is divided among the states based on their FY2012 shares of the whole. However, for FY2014, should any state receive less than 95 cents for every dollar contributed the highway account of the HTF the state's share is raised to that level. Given the excess of federal highway spending over HTF revenues in recent years, it is unlikely that any adjustments will have to be made.

Second, an amount for each state's apportionments from the Metropolitan Planning and CMAQ programs is set aside from each state's initial amount, based on the relative size of the state apportionments for FY2009 for these programs.

Third, the remainder of each state's "initial amount" is divided among the three remaining core programs as follows: 63.7% is apportioned to the National Highway Performance Program, 29.3% to the Surface Transportation Program, 7% to the Highway Safety Improvement Program ($220 million annually is set aside from HSIP for the Rail-Highways Crossings Program).

The Transportation Alternatives (TA) Program is then funded via a series of setasides at a level of 2% of all MAP-21 authorized federal-aid highway and highway research funds. Each state's core formula program and Metropolitan Planning apportionments are reduced proportionally to fund each state's TA. **Table 2** shows the dollar amounts of the aggregate programmatic split.

Core Highway Formula Programs

MAP-21 reduces the number of discrete funding programs by two-thirds to roughly 30 programs. Most of this reduction is accomplished by absorbing formerly separate activities and eligibilities into the new core programs discussed below. The core programs also have many areas of overlapping eligibility. Under MAP-21, the five core programs plus metropolitan transportation planning are authorized at $37.477 billion for FY2013 and $37.798 billion for FY2014.

Table 2. Apportioned Programs (Contract Authority) (millions of dollars)

Program	FY2013	FY2014	Total
National Highway Performance Program (NHPP)	21,752	21,936	43,687
Surface Transportation Program (STP)	10,005	10,090	20,095
Highway Safety Improvement Program (HSIP)	2,390	2,411	4,801
Congestion Mitigation & Air Quality Improvement Program	2,209	2,228	4,437
(CMAQ)			
Metropolitan Transportation Planning	312	314	626
Transportation Alternatives (TA)	809	820	1,629
Total	37,477	37,798	75,275

Source: Federal Highway Administration. The MAP-21 programmatic split is estimated. $220 million is set aside annually for Railway-Highway Crossings from HSIP.

National Highway Performance Program (NHPP; §1106)

NHPP has become the largest of the restructured federal-aid highway programs, with authorizations of $21.8 billion for FY2013 and $21.9 billion for FY2014. The program supports improvement of the condition and performance of the National Highway System,[9] combining the former Interstate Maintenance Program, the National Highway System Program, and the Highway Bridge Program's on-system component. NHPP includes projects to achieve national performance goals for improving infrastructure condition, safety, mobility, or freight movement, consistent with state or metropolitan planning; construction, reconstruction, or operational improvement of highway segments; construction, replacement, rehabilitation, and preservation of bridges, tunnels, and ferry boats and ferry facilities; inspection costs and the training of inspection personnel for bridges and tunnels; bicycle transportation infrastructure and pedestrian walkways; intelligent transportation systems; and environmental restoration, as well as natural habitat and wetlands mitigation within NHS corridors. If Interstate System and NHS bridge conditions in a state fall below the minimum conditions established by the Secretary of Transportation, certain amounts of funds would be transferred from other specified programs in the state. NHPP funds may be used for Appalachian Development Highway System projects with no state match.

Surface Transportation Program (STP; §1108)

STP remains the federal-aid highway program with the broadest eligibility criteria. Funds can be used on any federal-aid highway, on bridge projects on any public road, on transit capital projects, on non-motorized paths, and on bridge and tunnel inspection and inspector training. MAP-21 authorized $10 billion for FY2013 and $10.1 billion for FY2014. Although Transportation Enhancements are funded under the new Transportation Alternatives program, these types of projects can also be funded under STP if a state wishes. MAP-21 also allows STP funds to be used for Appalachian Development Highway System Projects with no state match.

Half of each state's STP funds are to be distributed within the state based on population. The remainder may be spent anywhere in the state. STP funds equal to 15% of the state's highway bridge apportionment for FY2009 are to be set aside for off-system bridges. MAP-21 included a special rule allowing some STP funds reserved for rural areas to be used on minor collector roads.

Highway Safety Improvement Program (HSIP; §1112)

HSIP remains largely as it was under SAFETEA, supporting projects that improve the safety of road infrastructure by correcting hazardous road locations, such as dangerous intersections, or making road improvements such as adding rumble strips. HSIP is funded at $2.39 billion for FY2013 and at $2.41 billion for FY2014. The Rail-Highway Grade Crossing Program was continued through a $220 million annual setaside.

Congestion Mitigation and Air Quality Program (CMAQ; Section 1113)

Under Map-21, CMAQ is authorized at roughly $2.209 billion for FY2013 and $2.411 billion for FY2014. Eligibility was expanded to include demand-shifting projects such as telecommuting, ridesharing, and road pricing.

Transportation Alternatives Program (TA; §1122): Transportation Enhancements/Non-Motorized Transportation Alternatives

In MAP-21, Congress changed the Transportation Enhancements program and other non-motorized transportation programs,[10] compromising between the positions of groups that wanted more funding for these programs and groups that wanted to eliminate these programs entirely.

The compromise eliminated certain types of activities from the list of eligible transportation enhancements,[11] renamed the transportation enhancements group of activities "transportation alternatives," and combined this group of activities with the former Recreational Trails and Safe Routes to

Schools programs under one umbrella program called Transportation Alternatives (TA). TA fund also may be used for "planning, designing, or constructing boulevards and other roadways largely in the right-of-way of former Interstate System routes or other divided highways." TA is a set-aside from each state's NHPP, STP, HSIP, CMAQ, and Metropolitan Planning apportionments amounting to roughly 2% of total highway funding. The amount available to each state is equal to the amount the state was required to set aside for Transportation Enhancements in FY2009. MAP-21 reduced the total amount set aside for these programs, from $1.2 billion in FY2011 to $809 million in FY2013 and $820 million in FY2014.[12]

There is no specific funding level for any of the programs within this group.[13] States are required to allocate 50% of the funds to local entities for obligation. If states do not obligate the remaining 50% of funding, they then may use these funds for any TA- or CMAQ-eligible projects once the unobligated amount accumulates to 100% of the state's annual TA set-aside.

MAP-21 also makes bicycle facilities and pedestrian walkways eligible expenses under the National Highway Performance Program, the Surface Transportation Program; and the Highway Safety Improvement Program.

Transferability among the Core Programs (§1509)

MAP-21 permits states to transfer up to 50% of any apportionment to any other apportionment program. However, no transfers are permitted of funds that are suballocated to areas by population (such as STP) or of Metropolitan Planning funds.

Other Highway Programs

Emergency Relief (ER) Program

Section 1107 clarifies eligibility criteria regarding roads and bridges damaged by natural disasters or catastrophic failures from an external cause. Roads already closed to traffic or already scheduled for the construction phase in the approved statewide transportation improvement plan at the time of the disaster are not eligible for ER funds. It also reiterates that ER funds can only be used on federal-aid highways. Within two years of a natural disaster or catastrophic failure, the state must apply and provide a comprehensive list of all eligible project sites and repair costs. The total cost of an ER project may not exceed the cost of repair or reconstruction of a comparable facility. The U.S. territories may not receive more than $20 million in a single fiscal year.

Section 1508 allows the 180-day emergency period during which the federal government pays 100% of repair costs to be adjusted for time lost due to lack of access to damaged facilities. Also, 90% federal share may be allowed at the discretion of the Secretary of Transportation if the cost to repair exceeds the annual state apportionment under 23 U.S.C. 104.[14] FEMA, not FHWA, will fund debris removal for major disasters declared under the Stafford Act. In the past, ER funds were often used for first-pass debris removal on disaster-impacted federal-aid highways.

Territorial and Puerto Rico Highway Program

Section 1114 funds the Puerto Rico and Territorial Highway programs at $150 million and $40 million annually, respectively, for both FY2013 and FY2014.

Appalachian Development Highway System Program (ADHS)

MAP-21 eliminates the ADHS as a freestanding program, incorporates its eligibilities into NHPP and STP, and provides for a 100% federal share for ADHS projects to encourage spending from these programs to complete the system. Within one year of enactment, each state represented on the Appalachian Regional Commission is to establish a plan for completion of the designated corridors within the state, including performance targets and a target completion date. In addition, to encourage states with significant uncompleted ADHS route miles to maintain their efforts toward completion, MAP-21 requires that if a state's estimated cost to complete its unfinished portion of the ADHS exceeds 15% of the total cost to complete of the entire system's unfinished mileage, then the state's completion plan may not reduce the state's obligated funds for the ADHS for any subsequent fiscal year.

Projects of National and Regional Significance (PNRS)

Section 1120 establishes a program similar to the program of the same name in SAFETEA. Budget authority, not contract authority, of $500 million is provided for FY2013. This discretionary program would require an appropriation before funds could be made available. The purpose is to fund critical high-cost surface transportation infrastructure projects that are difficult to complete with existing funding but would generate national and regional economic benefits, increase global competitiveness, reduce congestion, improve roadways vital to national energy security, improve the movement of freight and people, and improve transportation safety.

Construction of Ferry Boats and Ferry Terminal Facilities

Section 1121 changes the existing Title 23 discretionary Ferry Boats and Ferry Terminal Facilities Program into a formula program that includes no set-asides for specific states, as became the practice under the discretionary program.[15] MAP-21 provides the program with $67 million annually for FY2013 and FY2014. The funding is to be apportioned according to this formula: 20% based on a ferry system's total passengers for the most recent fiscal year relative to the number of passengers carried by all ferry systems; 45% based on the number of vehicles carried per day by the system relative to the number of vehicles carried by all systems; and 35% based on the total route miles serviced by the ferry system relative to the total route miles serviced by all ferry systems. Ferry boats and ferry facilities would also be eligible for formula funds under the National Highway Performance Program.

Federal Lands and Tribal Transportation Programs

Section 1119 restructures several programs to create the Federal Lands and Tribal Transportation Program. The new program has three main components:

- the Tribal Transportation Program, authorized at $450 million annually, replaces the Indian Reservation Roads Program;
- the Federal Lands Transportation Program, authorized at $300 million annually, combines the Park Roads and Parkways Program and the Refuge Roads Program;
- the Federal Lands Access Program, authorized at $250 million annually, replaces the Public Lands Highways Program.

The Tribal Transportation Program uses a new statutory formula for distributing some of the funds among tribes. This formula, to be phased in over four years beginning in FY2013, is based partly on road mileage and tribal population and partly on the relative need and population factors used under SAFETEA. MAP-21 also authorizes the Tribal High Priority Projects Program to fund a tribe's most important projects that cannot be completed with a regular annual allocation or for a project which is the result of an emergency or disaster. This is a new discretionary program, but compares with a setaside from the former Indian Reservation Roads Program. MAP-21 authorizes $30 million from the general fund for this program for FY2013 and FY2014.

Funding for the Federal Lands Access Program is allocated among the states by a formula that takes into account the amount of federal land, the number of recreational visitors, the number of miles of federal roads, and the number of federally owned bridges. From the funding for the Federal Lands Transportation Program $240 million is made available to the National Park Service and $30 million to the Fish and Wildlife Service. The remaining funding will be allocated among three other federal land management agencies, the Forest Service, the Corps of Engineers, and the Bureau of Land Management.

TOLLING

Law previous to MAP-21 allowed tolling of non-Interstate federal-aid highways. However, toll-free Interstate Highways, other than bridges or tunnels, could be converted to toll roads only under a pilot program limited to three projects. No states have applied to use tolls to fund new Interstate routes under the Interstate System Construction Toll Pilot Program, but some states are adding express toll lanes to existing roads and imposing congestion pricing under the Express Lanes Demonstration Program and the Value Pricing Pilot Program.

MAP-21 allows for the construction of totally new Interstate Highways as toll roads, essentially mainstreaming the Interstate System Construction Toll Pilot Program. It also allows for the addition of toll lanes on an existing Interstate Highway as long as the total number of "free" lanes is not reduced. The act eliminates the long-standing requirement that a toll agreement be executed with the FHWA prior to tolling a facility under the mainstream tolling programs. The act requires that by October 1, 2016, all federal-aid highway toll facilities implement interoperable tolling technology.

Tolls are important revenue streams for many public private partnerships and alternative financing mechanisms. The continued support in Congress for using congestion pricing fits well with expanded tolling of HOV and express lanes as well as the now allowable tolling of extended Interstate routes. Congress, on the other hand, did not expand the Interstate Reconstruction and Rehabilitation Pilot Program following public statements by participants that they were hoping to direct as much of the burden of tolls to interstate travelers as possible.

Transportation Infrastructure Finance and Innovation Act (TIFIA) Program

The TIFIA program provides secured loans, loan guarantees, and lines of credit for major surface transportation projects. Loans must be repaid with a dedicated revenue stream, typically a project-related user fee. MAP-21 greatly enlarges TIFIA by increasing its funding from $122 million annually to $750 million in FY2013 and $1 billion in FY2014. DOT estimates that after administrative costs and application of the obligation limitation it will have $690 million for credit subsidy support in FY2013 and $920 million in FY2014.[16] Assuming an average subsidy cost of 10%, this may provide DOT with the capacity to lend $6.9 billion in FY2013 and $9.2 billion in FY2014.[17] MAP-21 also increases the maximum share of project costs that TIFIA may provide from 33% to 49%, probably lowering the share of nonfederal resources leveraged with federal loans.

Another change made by MAP-21 is permitting TIFIA credit assistance to be provided for a program of projects secured by a common security pledge. This would be accomplished through a "master credit agreement." Prior to MAP-21, TIFIA only allowed agreements on a project-byproject basis. The Los Angeles County Metropolitan Transportation Authority (Metro), for one, has sought this change to accelerate the financing of 12 transit projects (known as the 30/10 Initiative).[18] The master credit agreement also establishes a way to make a commitment of future credit assistance contingent on the availability of funds.

The minimum amount of TIFIA assistance for a single project remains $50 million, or $15 million for intelligent transportation system projects, except that MAP-21 includes a threshold of $25 million for rural infrastructure projects.[19] MAP-21 also sets aside 10% of program funds to assist rural projects. Additionally, whereas loans for urban projects must be charged interest not less than the Treasury rate, rural projects that are assisted by the rural setaside are to be offered loans at half the Treasury rate. Rural projects are defined very expansively to include any project in an area other than a city with 250,000 or more inhabitants.

Prior to MAP-21, projects seeking TIFIA assistance were evaluated by DOT on eight criteria.[20] MAP-21 eliminates these selection criteria and now permits TIFIA assistance for any eligible project. One of the key eligibility criteria is creditworthiness. To be eligible, a project's senior debt obligations and the borrower's ability to repay the federal credit instrument must receive an investment-grade rating from at least one nationally recognized credit rating agency. The TIFIA assistance must also be determined to have several beneficial effects: fostering a public-private partnership, if appropriate; enabling the project to proceed more quickly; and reducing the contribution of federal grant funding. Other eligibility criteria include satisfying planning and environmental review requirements and being ready to contract out construction within 90 days after the obligation of assistance. Applications for assistance must be accepted by DOT on a rolling basis.

Changes to the TIFIA program, particularly the big increase in funding, have generally been well received. Nevertheless, there is some concern that the increase in TIFIA's share of project costs to 49% will reduce the non-federal share of project costs and possibly also "crowd out" private financing. Although DOT's ability to administer quickly such a large funding boost is also a concern, the agency acted to begin ramping up credit assistance by issuing a notice of funding availability just several weeks after MAP-21's enactment.

MAP-21 does not contain other major alternative financing provisions, such as federal funds to capitalize state infrastructure banks or increased volume limits on private activity bonds. Nor were there provisions to create a national infrastructure bank or to reinstate Build America Bonds.

TRANSIT

The public transit provisions of MAP-21 are contained in Division B, the Federal Public Transportation Act (FPTA) of 2012. Like MAP-21 as a whole, the FPTA authorizes funds for the remainder of FY2012 and for FY2013 and FY2014. Moreover, like MAP-21 as a whole, the transit provisions simplify the structure of the public transit program, eliminate discretionary programs in favor of formula programs, and introduce performance management.

Funding

For federal transit programs the bill authorizes $10.578 billion in FY2013 and $10.695 billion in FY2014, slight increases in nominal terms from the $10.458 billion in FY2012. About 80% of the authorized funds are from the mass transit account of the highway trust fund, with the remaining 20% authorized to come from the general fund (**Table 3**).

Program Restructuring

FPTA restructures the federal transit program. The Fixed Guideway Modernization Program is replaced with a new State of Good Repair (SGR) Grant Program. Funding for the SGR Program is increased to $2.136 billion in FY2013 and $2.166 billion in FY2014 from the $1.667 billion allotted to Fixed Guideway Modernization Program in FY2012. The new State of Good Repair program, funded from the mass transit account of the highway trust fund, has two components:

- The **High Intensity Fixed Guideway SGR Program** distributes 97.15% of the funding by formula for maintaining fixed guideway transit systems in a state of good repair. Funding is distributed by a new formula that uses fixed guideway vehicle miles and route miles that are at least seven years old.
- The **High Intensity Motorbus SGR** program distributes the remaining 2.85% of the funds by formula for public transportation provided on a high occupancy vehicle (HOV) facility. Funding is distributed by a formula that uses high-intensity motor bus vehicle miles and route miles for revenue services that are at least seven years old.

Linked with the intentions of the SGR program, transit agencies are required by FPTA to develop an asset management system, including an asset management plan with performance targets based on performance metrics developed by the Secretary. The Bus and Bus Facilities Program is retained in MAP-21, but at a much reduced funding level. Funding is authorized at $422 million in FY2013 and $428 million in FY2014, down from $984 million in FY2012. Additionally, funding from this heavily earmarked discretionary program is now to be distributed by formula with each state and territory receiving a minimum allocation and the remaining funds distributed according to population and service levels. A provision passed by the House Transportation and Infrastructure Committee to distribute bus program funding to providers of bus transit in urbanized areas that do not offer rail services was not adopted. Another significant change in MAP-21 is the combination of the Elderly Individuals and Individuals with Disabilities Program and the New Freedom Program, which provides formula funding for the disabled, into a single program. The new combined program is called the Enhanced Mobility of Seniors and Individuals with Disabilities Program. This program is authorized at $255 million in FY2013 and $258 million in FY2014 compared with $226 million for the Elderly and Disabled and New Freedom programs combined in FY2012. Funds in the new combined program are distributed by formula in the way that New Freedom funding was distributed: 60% is apportioned to large urbanized areas, 20% to small urbanized areas, and 20% to rural areas. Apportionments to specific areas are based on the number of elderly and disabled residents. Requirements for a locally developed, coordinated human services transportation plan are maintained.

Table 3. Public Transit Funding Authorized by MAP-21 (thousands of dollars)

Program	FY2013	FY2014
Total	10,578,000	10,695,000
Trust Funded Programs	8,478,000	8,595,000
Urbanized Area Formula Program	4,397,950	4,458,650
Passenger Ferry Boat Program (discretionary)	*30,000*	*30,000*
Operational Support of State Safety Oversight	*21,990*	*22,293*
State of Good Repair	2,136,300	2,165,900
High Intensity Fixed Guideway	*2,075,415*	*2,104,172*
High Intensity Motorbus	*60,885*	*61,728*
Rural Area Formula Program	599,500	607,800
Public Transportation on Indian Reservations	*30,000*	*30,000*
Appalachian Development Public Transportation	*20,000*	*20,000*
Rural Transportation Assistance Program (RTAP)	*11,990*	*12,156*
Projects of National Scope	*1,799*	*1,823*
Growing States and High Density States Formula	518,700	525,900
Bus and Bus Facilities (now §5339)	422,000	427,800
Mobility of Seniors and Individuals with Disabilities	254,800	258,300
Planning	126,900	128,800
Pilot Program for Transit Oriented Development	10,000	10,000
National Transit Institute	5,000	5,000
National Transit Database	3,850	3,850
Bus Testing Facility	3,000	3,000
General Funded Programs	2,100,000	2,100,000
New Starts	1,907,000	1,907,000
FTA Administration	104,000	104,000
Research, Development, Demonstration, Deployment	70,000	70,000
Low and no emissions buses	*45,500*	*45,500*
Low/no emissions bus facilities and equipment	*7,000*	*7,000*
Transit Cooperative Research Program (TCRP)	7,000	7,000
Technical Assistance and Standards Development	7,000	7,000
Human Resources and Training	5,000	5,000
Emergency Relief Program	ssaan	ssaan

Source: Federal Transit Administration, MAP-21 Fact Sheet: Funding Summary.
Notes: ssaan = such sums as are necessary.

The Jobs Access and Reverse Commute (JARC) formula program is eliminated in MAP-21, but the activities carried out under this program are made an eligible expense under the Urbanized and Rural Area Formula programs. MAP-21 also eliminates several discretionary programs including the Clean Fuels Grant Program, the Transit in Parks Program, the Over-the-Road Bus Program, and the Alternatives Analysis Program.

The Growing States and High Density State Formula Program is maintained at a somewhat higher funding level. Funding distributed through this formula is authorized at $519 million in FY2013 and $526 million in FY2014, up from $465 million in FY2012.

The Urbanized Area Formula Program is maintained largely as it was before MAP-21, but at an increased funding level. Funding is authorized at $4.398 billion in FY2013 and $4.459 billion in FY2014, up from $4.160 billion in FY2012. The Small Transit Intensive City (STIC) setaside from the Urbanized Areas Formula Program is increased from 1% to 1.5%. In addition, 0.5% is set aside for the state safety oversight program grants. MAP-21 also creates a new discretionary ferry boat grants program with $30 million set aside from the Urbanized Area Formula Program.

Similarly, the Rural Area Formula Program is maintained (although the name is changed from the Non-Urbanized Area Formula Program) and funding is increased from $465 million in FY2012 to $600 million in FY2013 and $608 million in FY2014. The formula used to apportion Rural Area program funds that has traditionally been based on rural land area and population is changed in MAP-21 to incorporate vehicle revenue miles and low-income individuals. Setasides from the rural formula program for the Rural Transit Assistance Program (RTAP) and the Public Transportation on Indian Reservations Program are maintained, and a new $20 million setaside for Appalachian Development Public Transportation Assistance is added. This new program mirrors the existing highway program for Appalachian development. MAP-21 doubles the amount set aside for Public Transportation on Indian Reservations from $15 million to $30 million annually. Of the $30 million, $25 million is distributed by formula and $5 million competitively.

MAP-21 creates the Public Transportation Emergency Relief Program that, like the Appalachian Development Public Transportation Assistance Program, mirrors an existing highway program. The emergency relief program, akin to the existing Highway Emergency Relief Program, would provide funding for capital and operating costs in the event of a natural or man-made disaster. The bill authorizes such sums as may be necessary to carry out this new program.

MAP-21 did not include proposals to increase the federal share of bus and bus facilities capital projects for transit systems that contract out at least 20% of fixed route bus service. However, MAP-21 does direct the Secretary of Transportation to promote private sector participation in public transit by providing technical assistance and education to transit agencies and by identifying impediments to public-private partnerships.

New Starts Program

MAP-21 reauthorizes the New Starts Program that provides funding to build new fixed-guideway transit systems and add to existing systems. Funding for the New Starts program in MAP-21 is authorized at $1.907 billion for both FY2013 and FY2014, a slightly reduced level from the $1.955 billion authorized in FY2012. As was previously the case, funding is authorized from the general fund.

MAP-21 makes substantial changes to the New Starts program. To begin with, the bill allows New Starts program funds to be used for substantial investments in *existing* fixed guideway systems that increase the capacity of a corridor by at least 10%. These types of projects are termed "core capacity improvement projects." It also authorizes the evaluation and funding of a program of interrelated projects.

MAP-21 simplifies the New Starts process by reducing the number of major stages from four to three. The new stages are termed project development, engineering, and construction.[21] To enter the project development phase, the applicant now only needs to apply to FTA and initiate the NEPA process. The bill eliminates the alternatives analysis that is separate from the alternatives analysis required by NEPA. Along with the NEPA work, during project development the project sponsor must develop the information needed by FTA to review the justification and the local financial commitment. Generally, the applicant has two years to complete project development. MAP-21 requires FTA to use an expedited process to review a sponsor's technical capacity if it has successfully completed a fixed guideway or core capacity project in the recent past.

The project is permitted to enter into the engineering phase once the NEPA process is concluded, the project is selected as the locally preferred alternative, the project is adopted into the metropolitan plan, and the project is justified on its merits. After engineering, a project is eligible to enter into a full funding grant agreement with FTA and to move into the construction phase.

MAP-21 also tries to advance projects more quickly using special warrants for projects of which the federal share is $100 million or less or 50% or less of the total project cost. According to FTA special warrants are "ways in which projects may qualify for automatic ratings on the project justification criteria."[22] For Small Starts projects, those requesting $75 million or less in federal assistance and costing in total $250 million or less, there are just two phases, project development and construction. Small Starts are also defined to include corridor-based bus rapid transit, a rail-like service that does for the most part operate in a separate right of way. Unlike SAFETEA, which reserved $200 million of the overall program authorization for Small Starts, MAP-21 does not reserve funds for Small Starts projects in FY2013 and FY2014. This may change the mix of New Starts and Small Starts projects that are funded, particularly because the newly permitted grants for core capacity projects could reduce the amount of funding available for other projects.

MAP-21 creates a pilot program for expediting New Starts project delivery, limited to three projects. A common criticism of New Starts is that it takes too long to develop and deliver projects, a criticism that MAP-21 seeks to address. An issue going forward, therefore, will be whether the changes made in MAP-21 do actually speed the process.

Operating Assistance

For the most part, MAP-21 maintains the prohibition on the use of federal funds for transit operating expenses in urbanized areas of 200,000 or more residents. However, it adds an exception to this general prohibition that is generally known as the "100 bus rule." MAP-21 permits transit systems in these larger urbanized areas operating 76 to 100 buses in peak service to use 50% of their Urbanized Area apportionment for operating expenses. For transit systems operating 75 or fewer buses in the peak period, the allowable amount is 75%.

Transportation Planning and Performance Management

Arguably the biggest change made by MAP-21 in transportation planning is a requirement for the use of performance management throughout the planning process. Performance management requires establishing performance measures and setting targets. MAP-21 makes changes in planning requirements at the national, state, and metropolitan levels.

National Planning

MAP-21 requires DOT to develop performance measures that address seven national goals: safety; infrastructure condition; congestion reduction; system reliability; freight movement and economic vitality; environmental sustainability; and reduced project delivery delays. In most cases, MAP-21 requires performance targets in support of these national goals to be set by states, metropolitan planning organizations (MPOs), regional planning organizations, and public transit agencies. MAP-21 does require DOT, however, to set minimum standards for states to use in bridge and pavement management systems on the National Highway System.

At the national level, MAP-21 also requires DOT to develop a national freight strategic plan. Among other things, the plan is required to include an assessment of the conditions and performance of the national freight network, forecasts of freight volumes over the next 20 years, identification of highway bottlenecks and the costs associated with addressing them, barriers to improving freight performance, and best practices for improving the national freight network.

Statewide Planning

In many respects, MAP-21 leaves state planning requirements as they were. Each state is still required to develop a statewide transportation plan and a statewide transportation improvement program. However, there are some changes (§1202). States are required to incorporate a performance-based approach into transportation planning in support of national goals. Performance targets are to be set by each state in coordination with MPOs. Each state must include in its statewide plan an evaluation of progress toward achieving its performance targets. Each state's transportation improvement program is also required to include a description of how its investment priorities will help achieve the stated performance targets.

As part of the new National Highway Performance Program (§1106), each state is required to develop a risk-based asset management plan for the

National Highway System that includes performance targets and an investment strategy. A state that fails to make significant progress toward achieving its targets has to submit a description of actions it will undertake to achieve them. As part of the planning, DOT must set minimum standards for the condition of Interstates and bridges on the National Highway System. If the condition of a state's bridges falls below that minimum, the state is required to redirect its federal apportionments to bring those facilities up to par. Additionally, if a state fails to adequately develop and implement a risk-based asset management plan, the maximum federal share of projects under the program is reduced to 65%. MAP-21 provides for states to create regional transportation planning organizations as planning entities in nonmetropolitan areas, somewhat akin to MPOs in urbanized areas.

Metropolitan Planning

MAP-21 did not include several proposed changes concerning MPOs, particularly alterations in population thresholds for various activities. Consequently, MAP-21 leaves metropolitan transportation planning requirements much as they were. MPOs are still required in urbanized areas, places of 50,000 or more residents, and these MPOs are still required to develop long-range plans and transportation improvement programs.

As with statewide planning, the biggest change in metropolitan planning is the requirement for a performance-based approach to support national goals. MAP-21 requires MPOs to establish performance targets that address the performance measures established by DOT. Each MPO must include in its plan an evaluation of the region's progress toward achieving its performance targets. The transportation improvement program must be designed to make progress toward the targets.

MPOs in transportation management areas with over 1 million residents that are in non-attainment or maintenance for air quality must also develop performance plans that link CMAQ funded projects with performance targets for emissions and congestion reduction. MAP-21 also includes provisions for the optional development by MPOs of multiple scenarios, sometimes known as blueprint planning.

In many respects MAP-21 initiates a shift to outcome-based planning. Implementation issues, including the development of goals and performance measures, will be likely to preoccupy many MPOs for the foreseeable future.

The requirements in MAP-21 contain little in the way of enforcement against those MPOs that fail to follow Congress's direction.

Non-environmental Provisions Accelerating Project Delivery

Outside of the environmental review process, MAP-21 makes two main changes to existing law in an attempt to speed project delivery. First, in Section 1303, MAP-21 adds specific authority for state DOTs to enter into construction manager/general contractor (CM/GC) contracts. According to FHWA, CM/GC contracts occupy a middle ground between the traditional design-bid-build construction method and the more innovative design-build method in which a single contractor is responsible for all the design and construction work.[23] With a CM/GC contract, a state DOT employs a general contractor to provide advice during the design phase. If agreement can be reached on price and other details, the same firm may then be employed to build the project. With intimate knowledge of the project, it is believed the contractor is able to enter into such an agreement and can begin construction tasks before the design work is complete, thereby accelerating the delivery of the project.

Second, MAP-21 would increase the federal funding share (normally 90% for Interstate Highway projects and 80% for other projects) by 5% on highway projects that use some method for accelerating project delivery (§1304). This applies to projects funded from the National Highway Performance Program, the Surface Transportation Program, and Metropolitan Planning (the increased federal share is limited to 10% of a state's apportionments under these programs). Examples of methods listed in MAP-21 that may speed project delivery include use of prefabricated bridge elements and design-build and CM/GC contracting methods.

Another provision (§1301 (b)(4)(c)) required DOT to establish a demonstration project to streamline the process of relocating households and businesses displaced by transportation projects by permitting lump-sum payments for acquisition and relocation if elected by the displaced occupant. Other proposed changes to the relocation provisions in current law,[24] including permitting advance payment of moving costs and a lump-sum payment for acquisition and relocation costs, were not enacted in MAP-21.

ACCELERATING TRANSPORTATION PROJECT DELIVERY

Transportation project development is initiated and completed largely at the local, tribal, or state level, with ultimate project approval at the federal level from FHWA or FTA. Generally, the stages include initial project planning, preliminary design/engineering and environmental review, final design and rights-of-way acquisition, construction, and facility operation and maintenance. Although issues that may delay project delivery can occur at any of phase of development, legislative efforts during the reauthorization debate focused on expediting overall project delivery by streamlining the environmental review process. [25]

For surface transportation projects approved for funding under federal-aid highway or public transportation programs, the environmental review process involves FHWA or FTA working with other federal and state or tribal agencies to ensure project compliance with all local, state, tribal and federal environmental requirements. This interdisciplinary undertaking is accomplished largely by documenting DOT's consideration of a proposed project's environmental impacts as required under the National Environmental Policy Act (NEPA, 42 U.S.C. §4321 et seq.).[26]

Procedures necessary to document compliance with NEPA were the primary focus of legislative efforts to expedite overall environmental compliance in MAP-21.[27] However, the act also includes provisions not entirely applicable to the NEPA process, but relevant to other processes likely to take place during the planning or preliminary design phase.

Requirements Applicable to the Environmental Review Process

Before a project can be approved for funding under DOT programs, FHWA or FTA must ensure that the project will comply with all state, tribal, and federal requirements—including those intended to consider, prevent, or minimize the project's impacts to the environment. [28] For a given surface transportation project, environmental requirements may apply to a project as a result of its impact on a resource that is subject to some level of protection under Titles 23 or 49 (e.g., DOT's prohibition on the approval of transportation projects that use parkland or historic sites[29]) or under other federal law (e.g., the National Historic Preservation Act, the Clean Water Act, or Endangered Specifies Act, to name a few). In addition to federal requirements, a project may be subject to various local, state or tribal laws or regulations.

Provisions in MAP-21 intended to accelerate project delivery focus primarily on elements of the NEPA compliance process, but may extend beyond NEPA. Generally, those provisions fall into one of the following categories:

- Statements of national policy and codification of existing policies, processes, or procedures;
- Amendments to selected requirements pertaining to the environmental review process established under SAFETEA;
- Directives to DOT to designate certain projects as categorical exclusions (CEs) from NEPA review, or to apply the CE determination process in a certain way; and
- Directives to DOT to establish programs and gather information.

On balance, a complex range of factors will affect the degree to which the changes enacted in MAP-21 will accelerate environmental reviews and ultimately project delivery. It is possible that they that might instead slow project delivery by removing mechanisms to coordinate the potentially complex environmental compliance process or by adding requirements to that process.

Statements of National Policy and Codification of Existing Requirements

Compliance with NEPA is demonstrated pursuant to regulations promulgated by both DOT and the Council on Environmental Quality (CEQ).[30] Several provisions intended to accelerate project delivery in MAP-21 codify regulations that are currently included in either CEQ or DOT regulations implementing NEPA. MAP-21 also includes several provisions that state national policies or priorities similar to procedures or practices those currently being implemented by DOT (under existing regulation or through programs such as DOT's Every Day Counts program).[31] Those provisions include:

- Section 1301, Declaration of policy and project delivery initiative.
- Section 1310, Integration of planning and environmental review.
- Section 1311, Development of programmatic mitigation plans.
- Section 1319, Accelerated decision making in environmental reviews.

- Section 1320, Memoranda of agency agreements for early coordination.

Amendments to Environmental Review Requirements Enacted under SAFETEA

Several provisions in MAP-21 amend requirements applicable to the environmental review process enacted under SAFETEA, mainly in Section 139 of Title 23, "Efficient environmental reviews for project decisionmaking" (hereinafter §139). Provisions that may lead to appreciable changes in the NEPA environmental review process are Sections 1306 and 1309.

Section 1306, "Accelerated Decisionmaking," amends Section 139(h), Issue Identification and Resolution, to create an issue resolution process that may be initiated by DOT to ensure that project deadlines are met. The process would apply to projects that require an environmental impact statement (EIS) under NEPA if an agency with jurisdiction over some project-related decision cannot provide "reasonable assurances" the agreed-upon deadlines will be met. It also establishes financial penalties that could be imposed on an agency that fails, within a certain time frame, to issue or deny a permit, license, or other approval required under any federal law. This section thus authorizes DOT to impose a penalty on an agency that is exercising authority under law within its, not DOT's, jurisdiction (e.g., the U.S. Army Corps of Engineers' obligation to issue a permit for projects affecting navigable waterways pursuant to the Clean Water Act). Also, requirements in Section 1306 are structured in a way that may actually delay certain approvals. If an agency faces a possible penalty unless it acts on an application by a certain date, the agency may have an incentive to simply deny the approval.

Section 1309 amends Section 139 to add "Enhanced Technical Assistance and Accelerated Project Completion." Under these provisions certain projects involved in EIS preparation may receive technical assistance from DOT to resolve outstanding issues and project delays. Technical assistance may include providing additional staff, training, and expertise or supplying onsite assistance.

Additional amendments to Section 139 are included in MAP-21 under Sections 1305, "Efficient environmental reviews for project decisionmaking," 1307, "Assistance to affected federal and state agencies," and 1308, "Limitations on claims." Those sections largely involve requirements that may not appreciably change the environmental review process because they involve minor changes to the process or include requirement being implemented by DOT under current regulations or programs. For example, under Section

1305(a), DOT is directed to promulgate regulations to allow for programmatic approaches to the environmental review process. Such approaches are already allowed, and encouraged, under CEQ regulations.[32] DOT has already developed several nationally applicable programmatic evaluations that states may use in completing the environmental review process for certain types of projects[33] and has entered into programmatic agreements with individual state DOTs to expedite the process for documenting and demonstrating compliance with various federal requirements.

MAP-21 also amends the following NEPA-related programs created pursuant to SAFETEA:

- **State assumption of responsibility for categorical exclusions** (23 U.S.C §326). As originally enacted under SAFETEA, states were allowed to assume federal authority for determining whether certain designated activities are CEs. To assume DOT authority in this area, a state was required to enter into a memorandum of understanding (MOU) setting forth the responsibilities to be assigned to that state.[34] MAP-21 Section 1312 amends Section 326 to specify that DOT shall not require states to forego project delivery methods that are "otherwise permissible for highway projects" as a condition of allowing states to assume responsibility for CE determinations.
- **Surface Transportation Project Delivery Pilot Program** (23 U.S.C. §327). As originally enacted under SAFETEA, the pilot program allowed DOT to assign, and five states (Oklahoma, California, Texas, Ohio, and Alaska) to assume, the federal responsibilities under NEPA for one or more highway projects in that state. MAP-21 Section 1313 amends Section 327 to make permanent the pilot program, to allow all states to participate, and to expand the program to include NEPA responsibilities with respect to one or more railroad, public transportation, or multimodal project.

Provisions Pertaining to Categorical Exclusions

Projects eligible for CEs are those that, based on past DOT experience with similar actions, normally do not *significantly* impact the environment.[35] MAP-21 includes provisions that are broadly intended to expand the potential universe of projects processed as CEs. The degree to which those requirements will accelerate the environmental review process for projects processed as CEs will likely depend largely on how much the new provisions differ from current DOT or state DOT practices.

Under Section 1318, "Programmatic agreements and additional categorical exclusions," DOT is directed to survey projects processed as CE since 2005 and state agencies to suggest new CEs and, based on its findings, add projects to the regulatory list of CEs. This survey may identify categories of projects that were not routinely identified as CEs. However, DOT previously could add to its list of CEs at any time, as it deemed necessary. Also under Section 1318, DOT must propose a rulemaking to move certain highway, bridge, and highway safety projects from the group of CEs listed as those that require appropriate documentation to be approved as a CE to the list of projects that normally require limited documentation.[36] The actions listed in MAP-21 involve a potentially broad range of projects and are those that DOT has identified, from past experience, as potentially involving unusual circumstances. Regardless of whether such an action is listed among those that normally require limited NEPA documentation, DOT's obligation under NEPA to ensure that a CE determination for an individual project is appropriate will not change.

MAP-21 also includes directives to DOT to designate the following broad categories of action as CEs and to promulgate regulations applicable to each designation:

- **Emergencies** (§1315). The repair or reconstruction of any road, highway, or bridge that is in operation or under construction when damaged under certain declared emergencies or disasters.
- **Projects within a right-of-way** (§1316). Any project located within an existing operational right-of-way.
- **Projects of limited federal assistance** (§1317). Projects that receive less than $5 million of federal funds or with a total estimated cost of not more than $30 million, if federal funds are less than 15% of total estimated project cost.

With regard to emergencies, projects that likely meet the criteria described under Section 1315 are currently identified as CEs and may be exempt from NEPA altogether.[37] For categories of actions that would be processed as CEs due to their locations within a right-of-way or receipt of federal funds below a certain threshold, several factors call into question whether such a designation would result in an appreciable change to the environmental review process. A specific project proposed under CE designations described in MAP-21 Sections 1316 and 1317 would likely already be included among actions processed as CEs.[38] If not, individual projects seeking CE approval under either designation would still have to demonstrate that they have no significant environmental impacts and involve no unusual circumstances that may require an environmental assessment or an EIS. Additionally, such a project would still be subject to additional state, tribal, or federal laws or regulations, including any requirements to obtain permits, evaluations, consultations, or approvals. As a result, explicitly designating such actions as CEs may have little effect on either the NEPA or the broader environmental compliance requirements applicable to such projects.

Directives to DOT and GAO

MAP-21 also includes directives to DOT to establish certain programs and to DOT or GAO to gather certain information and report to Congress. To meet the directives, state DOTs may be required to gather data that may be burdensome, if not impossible, to collect or may yield results that do little to accelerate project delivery. Those directives to DOT include

- **Section 1321, Environmental Procedures Initiative.** DOT is required to establish an initiative to "review and develop consistent procedures for environmental permitting and procurement requirements." With regard to environmental permitting, procedures applicable to such requirements would be established by state or tribal agencies or federal agencies outside DOT (no environmental requirements under Title 23 involve permitting). DOT would have limited authority to develop procedures applicable to those requirements. Beyond reviewing potentially applicable permitting requirements, DOT activity to respond to this directive may be limited to issuing guidance applicable to permitting procedures (which DOT has already done for federal permitting requirements). It is unclear how procurement requirements may be related to environmental procedures.
- **Section 1323. Review of federal project and program delivery.** This section requires DOT to compare the completion times for CEs, EAs, and EISs initiated after 2005 to those prepared during a period prior to calendar year 2005; and to compare completion times for CEs, EAs, and EISs initiated from January 1, 2005 to July 6, 2012 to those initiated after July 6, 2012. DOT must report this information to Congress within one year after enactment. Data necessary to make

such comparisons are not easily available. State DOTs generally do not measure the time it takes to complete individual stages of project development or extract data specific to the time it takes to complete NEPA documents. Data may be available for EISs, but not for the vast majority of surface transportation projects approved with a CE or EA. For future projects, a requirement to track and measure the time to complete NEPA documents may add a potentially burdensome level of bureaucracy to projects otherwise subject to limited documentation and reporting to DOT. Also, if such information could be collected, the completion times for NEPA documents may reveal little about efforts to expedite the environmental reviews, since the review process may start, stop, and restart for a number of factors unrelated to NEPA.

Provisions in Section 1322, Review of State Environmental Reviews and Approvals for the Purpose of Eliminating Duplication of Environmental Reviews, direct GAO to

- review state laws and procedures for conducting environmental reviews for surface transportation projects and identify states that have environmental laws that provide environmental protections and opportunities for public involvement that are equivalent to those provided by federal environmental laws;
- determine the frequency and cost of carrying out federal environmental reviews duplicative of state reviews; and
- report to Congress within two years describing the results of the review and determination.

This directive reflects some misperceptions regarding the federal environmental review process as implemented pursuant to NEPA. State environmental review programs are not necessarily identical to the federal NEPA process. However, pursuant to NEPA regulations, any analysis, review, or consultation prepared for a state environmental review may be incorporated by reference into the necessary NEPA document. Further, the NEPA compliance process forms the framework under which *any* potentially duplicative state, tribal, or federal requirements applicable to a project, not just those related to environmental reviews, are identified. If a state frequently carries out federal environmental reviews duplicative of state reviews, such duplication of effort could indicate a misunderstanding of the compliance process required under DOT's NEPA regulations. Additionally, the directive to determine costs associated with such efforts imposes a data-collection burden on state DOTs.

State DOTs generally do not track costs of individual components of project development, including elements of the environmental review process. It may, therefore, prove difficult for states to 1) identify specific elements of the federal environmental review process that are duplicative of state requirements and 2) to extract project costs explicitly related to such a duplication of effort.

AMENDMENTS TO THE CMAQ PROGRAM

The Congestion Mitigation and Air Quality Improvement (CMAQ) Program was established to fund projects and programs that may reduce emissions from transportation-related pollutants, particularly in nonattainment and maintenance areas.

Pursuant to the Clean Air Act, the Environmental Protection Agency (EPA) has set ambient air quality standards for certain pollutants. Of relevance to transportation planning agencies are the National Ambient Air Quality Standards (NAAQS) for ozone, carbon monoxide, and particulate matter (distinguished as coarse and fine particulate, referred to as PM10 and PM2.5, respectively).[39] A geographic area that meets or exceeds NAAQS for a criteria pollutant is considered to be in "attainment"; an area that does not meet a standard is in "nonattainment."

A "maintenance" area is one that was previously in nonattainment, but is currently attaining the NAAQS subject to a maintenance plan.[40]

MAP-21 Section 1113 amends the following CMAQ program requirements:

- **Eligible Projects (§149(b)).** Explicitly includes the addition of turning lanes to the projects eligible for funding.
- **Special Rules (§149(c)).** Allows a state to obligate its CMAQ funds for a project in a nonattainment area regardless of the criteria pollutant the project aims to reduce; and specifies that CMAQ funds could be used for electric and natural gas vehicle infrastructure.
- **States Flexibility (§149(d)).** Specifies that a state with a nonattainment area may use a certain amount of its apportioned STP funds for projects eligible under the CMAQ program, determined in

relation to 2009 funding levels for projects funded under the CMAQ program.

- **Evaluation and Assessment of Projects (§149(i))**. Directs DOT to develop and maintain a database to disseminate information describing CMAQ projects and, in consultation with EPA, evaluate the cost-effectiveness of a range of projects. States and MPOs are required to consider this information in selecting projects or developing performance plans (as required under amended Section 149(l), discussed below).
- **Optional Programmatic Eligibility (§149(j))**. Allows technical assessment of selected program or projects, conducted at the discretion of MPOs, to demonstrate emissions reductions. That data could be used to show that similar projects meet CMAQ eligibility requirements.
- **Priority for Use of Funds in $PM_{2.5}$ Area (§149(k)).** In nonattainment or maintenance areas for $PM_{2.5}$, states and MPOs shall prioritize CMAQ fund distribution for projects proven to reduce those pollutants, including diesel retrofits.
- **§149(l) Performance Plan**. Requires MPOs to prepare performance plans for CMAQ-funded projects.
- **§149(m) Operational Assistance.** Allows states to obligate apportioned CMAQ funds for operating costs of transit projects or systems previously eligible for funding under the CMAQ program.

SAFETY PROGRAMS

National Highway Traffic Safety Administration (NHTSA)

NHTSA is the agency primarily responsible for highway safety. NHTSA promotes safety in two ways: through addressing driver behaviors that contribute to crashes (e.g., driving while intoxicated, speeding, and distracted driving), and through addressing the safety aspects of passenger vehicles (e.g., requiring seat belts, air bags, and electronic stability control).

Driver behavior is a state matter, not under federal control. Consequently, when Congress addresses driver behavior issues, it does so by encouraging states to act.

NHTSA's driver behavior programs are primarily grants to states to help pay for state actions addressing these issues. In MAP-21, Congress authorized essentially level funding for highway traffic safety grant programs: $670 million in FY2013, up 1% from $660 million in FY2012 (see Table 4).

Congress also made changes to NHTSA's safety grant program structure in MAP-21. It eliminated the state safety belt incentive grant program (most states had already qualified, and the one which had not was expected to qualify in the near future), consolidated two other occupant protection grant programs into one, and created two new grant programs:

- Distracted driving incentive grant program: to make grants to states that implement a law making distracted driving or texting while driving a primary offense (meaning that drivers can be stopped and ticketed for that offense). Thirty-five states already have laws making texting while driving a primary offense[41]; in the first year, DOT is allowed to use up to 25% of the funds to make grants to states that already have such laws or are not otherwise eligible.
- Graduated driver licensing incentive grant program: to make grants to states that implement graduated driver licensing laws that limit the use of cell phones, driving at night, and carrying passengers not related to the driver.

In MAP-21, Congress prohibited states from using any of their federal highway safety formula (popularly referred to as "Section 402") grant funding to purchase, operate, or maintain automated speed or red light cameras.

Congress also provided additional powers to DOT and increased penalties to address vehicle safety issues.

Federal Motor Carrier Safety Administration (FMCSA)

FMCSA is responsible for safety in the commercial motor vehicle industry; that is, generally, commercial interstate freight trucking and charter bus passenger travel. FMCSA promotes safety in two ways: by addressing commercial driver qualifications and activities, such as work hours, and by addressing the condition of commercial motor vehicles.

In MAP-21, Congress kept FMCSA's program structure roughly the same as under SAFETEA, and authorized roughly level funding for the agency's grant programs (see Table 5).

Table 4. NHTSA Highway Traffic Safety Grant Program Funding in MAP-21 (millions of dollars)

	2012	2013	2014
State Formula Grants (§402)	$235.0	$235.0	$235.0
Highway Safety Research & Development	105.5	110.5	113.5
(§403)			
National Safety Priorities (§405)		265.0	272.0
Occupant Protection Grants	80.5	42.4	43.5
State Traffic Safety Information System	34.5	38.4	39.4
Improvements Grants			
Impaired Driving Countermeasures	139.0	139.1	142.8
Grants Ignition Interlock Incentive Grant		20.9	21.42
Distracted Driving Grants	—	22.5	23.12
Motorcyclist Safety Grants	7.0	4.0	4.1
State Graduated Driver Licensing	—	13.3	13.6
Grants			
National Driver Register	4.8	5	5
High Visibility Enforcement Program	29.0	29	29
Administrative Expenses	25.3	25.5	25.5
Total	**$660.6**	**$670.0**	**$680.0**

Source: P.L. 112-441, §31101 & 31105; FY2012 figures from NHTSA's FY2013 Budget Estimates volume and from P.L. 112-55, 125 STAT. 657-658.

Notes: Due to MAP-21 program changes and additions, some FY2013 programs are not directly comparable to FY2012 programs. In the table, three separate occupant protection grant programs are consolidated for FY2012 for comparability to FY2013 & FY2014.

Table 5. FMCSA State Safety Grant Program Funding in MAP-21 (millions of dollars)

Program	FY2012	FY2013	FY2014
Motor Carrier Safety Assistance Grant Program	$212	$215	$218
Commercial Driver License Program Implementation Grant Program	30	30	30
Border Enforcement Grant Program	32	32	32
Commercial Vehicle Information Systems and Networks Grant Program	25	25	25
Safety Data Improvement Grant Program	3	3	3
New Entrant Audit Grant Program	29	32	32
Performance and Registration Information System Management Grant	5	5	5
Program			
Total	$336	$342	$345

Source: P.L. 112-141 §32603; FY2012 figures from FMCSA FY2013 Budget Estimates volume.

Congress added several new provisions intended to promote commercial motor vehicle safety. These include:

- a requirement that every commercial vehicle involved in interstate commerce and operated by a driver subject to the hours of service and record of duty requirements be equipped with an electronic onboard recorder (also known as a "black box") to improve compliance with hours of service regulations. This requirement will take effect around 2015;[42]
- increased enforcement penalties and powers for dealing with carriers operating without registration and "reincarnated" carriers (carriers which resume operation under a different name after having been shut down for safety violations);
- a directive to DOT to establish minimum entry-level training requirements for commercial drivers; and
- a directive to DOT to establish a national clearinghouse for controlled substance and alcohol test results of commercial drivers in order to improve compliance with commercial driver's license standards.

Motorcoach Safety

MAP-21 made several changes to the safety standards for motorcoaches, including requiring

- Electronic stability control systems to reduce the risk of rollover crashes;
- Increased roof strength to protect occupants in the event of rollover crashes; and
- Seat belts for passengers.

Congress also directed DOT to consider requiring:

- Tire pressure monitoring systems in order to reduce the risk of crashes caused by underinflated tires;
- Improved tire performance standards; and
- Stronger window standards to reduce the risk of passengers being thrown from motorcoaches during crashes.

Other Provisions

Congress directed DOT to establish accelerated licensing procedures to help veterans who received safety training and operated qualifying motor vehicles during their military service to obtain commercial driver's licenses.

FTA Transit Safety Oversight Program

The federal government has long regulated the safety of air travel, intercity passenger rail travel, interstate passenger bus travel, and interstate freight transportation by truck or rail. When Congress began to provide federal assistance for public transit, though, it specifically prohibited the federal government from regulating transit operations, including safety practices.[43] Public transit generally has an extremely low rate of fatalities, but the National Transportation Safety Board had identified weaknesses in transit agencies' safety practices and recommended federal regulation to improve the situation. In recent years, high-profile incidents on heavy-rail transit systems in Washington, DC and Chicago led to increased pressure for federal regulation.

In MAP-21, Congress directed FTA to create and implement a national public transportation safety plan, to include safety performance criteria for all modes of public transportation and minimum safety performance standards for public transportation vehicles not otherwise regulated by the federal government.

Congress also gave FTA the authority to set and enforce minimum safety standards for transit rail systems and to oversee state safety oversight programs. MAP-21 authorized $22 million annually for formula grants to eligible states' safety oversight programs to help pay for participation in the new safety program. This funding may be used for operational expenses, including employee training.

Freight Policy

Whether the federal government should make a more focused effort towards funding freight improvements was a major policy question leading up to the reauthorization debate. The Senate version of MAP-21 (S. 1813) created a separate funding program for freight transport but this was not enacted in the final bill. Instead, the local cost share for freight specific projects was reduced to 5% in the case of projects on the Interstate Highway System and 10% for other projects. MAP-21 enacted many planning provisions related to identifying infrastructure components most critical to freight transport.[44] This includes designation of a "primary freight network" (PFN) consisting of 27,000 centerline miles of existing roadways, based primarily on freight volume and in consultation with shippers and carriers (for comparison, the existing Interstate Highway System consists of approximately 47,000 centerline miles). The Secretary of Transportation can designate up to an additional 3,000 centerline miles of existing or planned roads as part of the PFN based on their future importance to freight movement. States can designate "critical rural freight corridors" based on the density of truck traffic if they connect the PFN or Interstate System with sufficiently busy freight terminals. A larger National Freight Network (NFN) will include the critical rural freight corridors, portions of the Interstate System not designated as parts of the PFN, and roads in the PFN. DOT, in consultation with partners and stakeholders, will develop a national freight strategic plan that will include identification of highway bottlenecks. DOT is required to report every two years on the condition and performance of the NFN.

Each state is encouraged, but not required, to create a state freight advisory committee made up of representatives of the freight community and a state freight plan "that provides a comprehensive plan for the immediate and long-range planning activities and investments of the State with respect to freight" (§1118). Among other things, a state's freight plan is to describe how it will improve the ability of the state to meet the national freight goals established by DOT.

Truck Size and Weight

No major changes to current truck size and weight provisions are included in MAP-21, but a new study and inventory of current state laws is required.[45] The act increases the weight allowance for truck idle reduction technologies from 400 lbs. to 550 lbs. and allows states to issue overweight permits to trucks carrying relief supplies if a major disaster is declared.[46]

Freight Broker Financial Security

The surety bond filing requirement for truck brokers and ocean freight forwarders was raised from $10,000 to $75,000 with the intention of fighting fraud in the industry.[47] Truck brokers and freight forwarders marry shippers (cargo owners) with carriers on an individual shipment basis. Small brokers contend this change is unnecessary and is being used by larger firms as a means of putting them out of business. Since brokers do not own any transportation assets, entry and exit in the industry is relatively easy. The bond is employed as a means of ensuring brokers have sufficient financial stake in the business and/or as a means of paying creditors when firms do go out of business.

Harbor Maintenance

Section 1536 of MAP-21 states as a non-binding Sense of Congress that the Administration should request and Congress should fully expend each year all of the revenues collected in the harbor maintenance trust fund for the operation and maintenance of the nation's federally maintained ports.[48] The provision also expresses the importance of protecting other critical Army Corps programs, including inland navigation, flood and coastal storm protection, and ecosystem restoration, from funding reductions.

The provision also directs the Administration to provide an annual estimate of national harbor maintenance needs, including an estimate of the percentage of waterways that will be available for use based on the annual budget request as well as how much funding would be needed to achieve 95% availability of the nation's ports and waterways within three years.

U.S. Flag Shipping Requirement Relaxed

Section 100124 reduces the percentage of U.S. food aid that must be shipped on U.S.-flag ships (which must be owned and crewed by U.S. citizens) from 75% to 50% and repeals the requirement that 25% of bagged or processed food aid be shipped through Great Lakes ports. The purpose of this so-called "cargo preference" law is to promote U.S.-flag shipping because the ships may be used to deploy military hardware and supplies during war or national emergency. Operating expenses under the U.S. flag are significantly higher than those under many foreign flags, which is why Congress supports U.S.-flag vessels through cargo preference, with the additional shipping cost paid by DOT's Maritime Administration. Section 100124 repeals a provision of the Food Security Act of 1985 (P.L. 99-198, subtitle C), which increased the cargo preference requirement from 50% to 75% of food aid tonnage.[49]

End Notes

[1] CRS Report R41869, The Donor-Donee State Issue in Highway Finance, by Robert S. Kirk.

[2] CRS Report R41947, Accelerating Highway and Transit Project Delivery: Issues and Options for Congress, by William J. Mallett and Linda Luther.

[3] A separate 0.1 cents per gallon tax on all fuels goes into the leaking underground storage tank (LUST) trust fund. LUST is administered by the Environmental Protection Agency and the states. It funds leaking underground storage tank cleanup and leak prevention activities. The authorization of this fund is not addressed in surface transportation legislation.

[4] In late FY2008, $8 billion was transferred to carry the highway account into the 2009 fiscal year (P.L. 110-318, September 15, 2008). In FY2009 the transfer was $7 billion (P.L. 111-46, August 7, 2009). The Surface Transportation Extension Act of 2010 (P.L. 111-148, March 18, 2010) transferred $14.7 billion more to the highway account. The third rescue package, P.L. 111-147, also transferred $4.8 billion to the mass transit account.

[5] Joint Committee on Taxation, Estimated General Fund and Trust Fund Effects of the Conference Agreement for the Revenue Provisions Contained in Division D of MAP-21, committee print, 112th Cong., 2nd sess., June 28, 2012, JCX58-12, pp. 1-3.

[6] Congressional Budget Office, Highway Trust Fund Projections: CBO August FY2012 Baseline 2011-2012, August 22, 2012. Because requests for reimbursement from the HTF may occur at any time and because treasury transfers to the HTF occur only twice each month and requests for reimbursement from the states can vary from month to month, FHWA deems it prudent to maintain a $4 billion minimum in the highway account to prevent having to delay payments to states due to insufficient funds. The equivalent prudent balance for the mass transit account is $1 billion.

[7] For a more detailed discussion see Federal Highway Administration, Financing Federal-Aid Highways, (Washington, 2007), pp. 9-10, http://www.fhwa.dot.gov/reports/financing federalaid/approp.htm#b.

[8] Ibid., pp. 19-22. To be contract authority the authorization must refer to Title 23, Chapter 1 of the U.S. Code, and it must be funded out of the highway trust fund.

[9] Section 1104 redefines the components of the National Highway System (NHS), which already included all Interstate highways and most major roads, allowing for the incorporation of additional principal arteries not previously included in the NHS and eliminating the national mileage limit on the designations. It also adds the strategic highway network to the NHS and sets forth clarified rules for modifications to the National Highway System and the Interstate System.

[10] §1122.

[11] Funding for transportation museums, scenic or historic highway programs, acquisition of scenic or historic easements and sites, and pedestrian and bicycle safety and education programs was eliminated.

[12] The new set-aside is equal to the amount set aside for transportation enhancements in FY2009. There is also a provision that allows states to use a portion of this funding for other purposes, under a particular circumstance: 50% of the funds are to be suballocated to local government authorities, and if the state does not use the remaining 50%, then in subsequent

years the state may use the amount in excess of one year's funding for any project eligible under CMAQ. As an example, if the Transportation Alternative set-aside for a state is $20 million, the state must suballocate $10 million to local governments, and then may use any funds in excess of the remaining $10 million balance for other CMAQ projects.

[13] Each state is required to obligate the same amount for recreational trails that it was apportioned for recreational trails in FY2009—but states are allowed to opt out of that requirement.

[14] Subsection (g) language appears to be written to benefit a particular project, group of projects, or area of roadways threatened by "continuous or frequent flooding" included within the "scope of a prior emergency declaration."

[15] The program is part of the Federal-Aid Highway Program because it is designed to permit federal participation in the construction of ferry boats and terminal facilities where it is not feasible to build a bridge, tunnel, or other normal highway structure in lieu of the use of a ferry.

[16] Department of Transportation, "Letters of Interest for Credit Assistance Under the Transportation Infrastructure Finance and Innovation Act (TIFIA) Program," 77 Federal Register 45411-45415, July 31, 2012, http://www.fhwa.dot.gov/ipd/pdfs/tifia/f y2013_tifia_nofa_073112.pdf.

[17] The subsidy cost is "the estimated long-term cost to the government of a direct loan or a loan guarantee, calculated on a net present value basis, excluding administrative costs," Federal Credit Reform Act of 1990 (FCRA), §502 (5A).

[18] Los Angeles County Metropolitan Transportation Authority (Metro), Metro's 30/10 Initiative, http://libraryarchives.metro.net/DB_Attachments/100524_30_10_Initiative.pdf.

[19] The law also provides eligibility for projects whose total expected costs are 33.3% of the amount of federal highway assistance apportioned in the most recent fiscal year to the state in which the project is located. This is unchanged in MAP-21.

[20] These were the amount of private participation; environmental impact; national or regional significance; project acceleration; creditworthiness; use of new technologies; reduced federal grant assistance; and consumption of budget authority.

[21] Previously, the New Starts process involved four major phases: planning and alternatives analysis; preliminary engineering; final design; and construction.

[22] Federal Transit Administration, Fact Sheet: Fixed Guideway Capital Investment Grants ("New Starts"), Section 5309, http://www.fta.dot.gov/documents/MAP-21_Fact_Sheet_-_Fixed_Guideway_Capital_Investment_Grants.pdf.

[23] Federal Highway Administration, "Every Day Counts Initiative: Accelerated Project Delivery Methods," http://www.fhwa.dot.gov/everydaycounts/projects/methods/index.cfm.

[24] Uniform Relocation Assistance and Real Property Acquisition Policies Act of 1970 (42 U.S.C. 4601 et seq.).

[25] For more information, see CRS Report R41947, Accelerating Highway and Transit Project Delivery: Issues and Options for Congress, by William J. Mallett and Linda Luther.

[26] CRS Report R42479, The Role of the Environmental Review Process in Federally Funded Highway Projects: Background and Issues for Congress, by Linda Luther.

[27] Under Title I, Federal-Aid Highways, Subtitle C.

[28] Requirements broadly referred to as "environmental" requirements may involve that arise as a result of the proposed project's impact to an affect community or to certain natural, cultural, or environmental resources.

[29] Pursuant to the "Preservation of parkland requirements" under 23 U.S.C. Section 138, more commonly referred to as "Section 4(f)."

[30] CEQ requirements at 40 C.F.R. Sections 1500-1508 apply broadly to all federal agencies and apply explicitly to NEPA compliance. DOT's regulations, "Environmental Impact and Related Procedures," build on the CEQ requirements as necessary to ensure DOT compliance with NEPA.

[31] For examples of DOT programs intended to streamline the NEPA and overall environmental review process, see information on FHWA's "Environmental Streamlining and Stewardship" at http://environment.fhwa.dot.gov/strmlng/ index.asp and DOT's "Every Day Counts" initiative at http://environment.fhwa.dot.gov/integ/edc.asp.

[32] See 40 C.F.R. §1508.29.

[33] For information regarding FHWA's national programmatic evaluations for transportation enhancements, see http://www.fhwa.dot.gov/environment/transportation_enhancements/ guidance/gmemo_program.cfm; for projects that require an evaluation under Section 4(f), see http://www.environment.fhwa.dot.gov/4f/4fnationwideevals.asp.

[34] Pursuant to directive included in SAFETEA, DOT established an MOU template and guidance to states that wanted to accept federal authority to process CEs, available online at http://www.fhwa.dot.gov/hep/6004moutemplate.htm.

[35] Such as actions that do not induce significant impacts to planned growth or land use for the area or have a significant impact on any natural, cultural, recreational, historic or other resources (23 C.F.R. §771.117(a)).

[36] Those projects are to be moved from 23 C.F.R. §771.117(d) to §771.117(c).

[37] "Emergency repairs under 23 U.S.C. §125" are explicitly listed as 23 C.F.R. §771.117(c)(9). Under 23 U.S.C. §125, "Emergency relief," a source of funding is provided for projects related to emergencies, as declared by a governor, and for disasters, as declared under the Stafford Act. Emergency relief projects are largely identical to the types of projects described in MAP-21 Section 1315. The Stafford Act waives NEPA procedures for certain federal actions taken or carried out within a presidentially declared emergency or disaster area. Regardless of their funding source or potential status as an "emergency," projects similar to those described in Section 1315 would likely meet the criteria applicable to actions already identified on by DOT as a CE, particularly bridge reconstruction or replacement and highway restoration, rehabilitation, or reconstruction.

[38] Such as bridge rehabilitation or reconstruction; highway resurfacing, restoration, rehabilitation, or reconstruction; construction of bus transfer facilities in a commercial area; or highway safety or traffic operations improvement projects.

[39] "Mobile sources," such as cars, trucks, buses, and other vehicles, are considered significant sources of these pollutants. NAAQS have also been established for lead and sulfur dioxide, but mobile sources are not significant sources of those pollutants.

[40] The Clean Air Act Amendments of 1990 directed EPA to establish the NAAQS. The following year, under provisions in the Intermodal Surface Transportation Efficiency Act of 1991 (ISTEA), the CMAQ program was established under 23 U.S.C. §149. The program was amended and program funding was reauthorized in both TEA-21 in 1998 and SAFETEA in 2005. For information about the program, see FHWA's "Air Quality" web page at http://www.fhwa.dot.gov/environment/air_quality/cmaq/.

[41] Governors Highway Safety Association, "Cell phone and texting laws," http://www.ghsa.org/ html/stateinfo/laws/cellphone_laws.html.

[42] §32301. The Secretary of DOT is required to prescribe regulations for the devices within one year after passage of MAP-21; the regulations shall apply to commercial motor vehicles beginning two years after the regulations are published as a final rule. The deadline to install the devices thus depends on when DOT completes the regulation.

[43] Commuter rail services, which operate on the national rail network, are subject to regulation.

[44] See §1115-1118.

[45] §32801 and §32802.

[46] §1510 and §1511.

[47] See §32918.

[48] For background information, see CRS Report R41042, Harbor Maintenance Trust Fund Expenditures, by John Frittelli, and CRS Report R41961, Army Corps Fiscal Challenges: Frequently Asked Questions, by Nicole T. Carter and Charles V. Stern.

[49] The origin of the U.S. flag shipping requirement for food aid cargo dates to two 1954 Acts: P.L. 83-480 and P.L. 83- 644.

In: Surface Transportation
Editor: Donna Neil
ISBN: 978-1-62417-431-5

Chapter 2

SURFACE TRANSPORTATION: FINANCING PROGRAM COULD BENEFIT FROM INCREASED PERFORMANCE FOCUS AND BETTER COMMUNICATION*

United States Government Accountability Office

ABBREVIATIONS

AASHTO	American Association of State Highway and Transportation Officials
ARTBA	American Road and Transportation Builders Association
DOT	department of transportation
FCRA	Federal Credit Reform Act
FHWA	Federal Highway Administration
FRA	Federal Railroad Administration`
FTA	Federal Transit Authority
GAN	Grant Anticipation Note
GARVEE	Grant Anticipation Revenue Vehicle
ITS	intelligent transportation system
LOI	letter of interest

* This is an edited, reformatted and augmented version of The United States Government Accountability Office publication, Report to Congressional Committees GAO-12-641, dated June 2012.

MAP-21	Moving Ahead for Progress in the 21st Century Act
NOFA	Notice of Funding Availability
OMB	Office of Management and Budget
OST	Office of the Secretary of Transportation
RRIF	Railroad Rehabilitation and Improvement Financing
SAFETEA-LU	Safe, Accountable, Flexible, Efficient Transportation Equity Act: A Legacy for Users
SANDAG	San Diego Association of Governments
SEP-15	Special Experimental Project Number 15
TIFIA	Transportation Infrastructure Finance and Innovation Act
TIGER	Transportation Investment Generating Economic Recovery

Why GAO Did This Study

Created in 1998, the TIFIA program is designed to fill market gaps and leverage substantial nonfederal investment by providing federal credit assistance to help finance surface transportation projects including highway, transit, rail, and intermodal projects. Since 2008, demand for the program has surged, annually exceeding budget resources for the program by a factor of more than 10 to 1. Given the increased demand and recent proposals to expand and modify the program, GAO was asked to review (1) the characteristics of TIFIA projects and how DOT tracks progress toward the program's goals, (2) the process DOT used to evaluate and select projects that submitted LOIs to apply for credit assistance in fiscal years 2010 and 2011, (3) the factors that affect project sponsors' decisions about whether to seek TIFIA credit assistance, and (4) the options proposed to modify the program. GAO reviewed laws and program guidance; interviewed DOT officials, project sponsors, and advisors involved in procuring credit assistance; and surveyed all state departments of transportation and other recent applicants about the TIFIA program.

What GAO Recommends

GAO recommends that DOT develop and use program performance measures to better assess progress in meeting TIFIA's goals and objectives. DOT should better disclose information on how it selects projects to apply for

TIFIA assistance through program guidance or other means to help ensure that the program is more transparent to Congress, applicants, and the public. DOT said it would consider the study's results.

What GAO Found

Projects that received credit assistance through the Transportation Infrastructure Finance and Innovation Act (TIFIA) program, administered by the Department of Transportation (DOT), tend to be large, high-cost highway projects. As of April 2012, DOT has executed 27 TIFIA credit agreements for 26 projects with project sponsors such as state DOTs and transit agencies. Overall, DOT has provided nearly $9.1 billion in credit assistance through 26 loans and one loan guarantee. By mode, there are 17 highway, 5 transit, and 4 intermodal projects.

Most projects have a total cost of over $1 billion. DOT monitors individual credit agreements but does not systematically assess whether its TIFIA portfolio as a whole is achieving the program's goals of leveraging federal funds and encouraging private co-investment. DOT has identified goals and objectives for the TIFIA program, but its limited use of performance measures makes it difficult to determine the degree to which the program is meeting these goals and objectives. Given that DOT already collects project data, it could use these data to better evaluate the program's overall progress toward meeting its goals.

In fiscal years 2010 and 2011, DOT used a competitive two-step process to evaluate and invite projects to apply for TIFIA credit assistance to address the considerable increase in demand for the program. First, a multimodal team scored and grouped letters of interest (LOI) using statutory criteria. Second, a group of senior DOT staff reviewed the LOIs based on the criteria and other factors, like available budget authority, and invited a subset to apply—the next step in securing TIFIA assistance. While recent applicants were satisfied with many aspects of the process, they also indicated, along with legal and financial advisors, that the selection process lacks transparency and creates uncertainty in their ability to implement projects. For example, some recent applicants told us it is difficult to understand what characteristics DOT uses to measure how well a project meets each criterion. DOT officials said the agency is taking steps to improve its evaluation process, but since many of the changes were initiated in 2012, it is too soon to tell if they will address recent applicants' concerns.

Several factors influence whether project sponsors seek TIFIA assistance. More than 30 of 36 recent applicants we surveyed cited TIFIA's repayment options (like deferring repayments for 5 years after project completion), low interest rate, and flexible structure (i.e., ability to subordinate TIFIA repayment) as important in their decision to seek assistance. To date, sponsors from 17 states have never sought TIFIA assistance. State DOT respondents from these states cited various reasons for this, including lack of eligible projects and state-imposed borrowing restrictions. Many of these state DOTs indicated that regardless of options for modifying the program, they have no plans to seek TIFIA assistance.

Several options to change the TIFIA program have been proposed by, among others, Congress and DOT; these options include increasing the program's funding, increasing the portion of costs that may be covered by TIFIA from 33 percent to 49 percent of project costs, and modifying the selection process. Each option has advantages and disadvantages and, if adopted, some could alter the original goals of the program—to leverage public funds and encourage private co-investment.

June 21, 2012

The Honorable Barbara Boxer
Chairman

The Honorable James M. Inhofe
Ranking Member
Committee on Environment and Public Works
United States Senate

The Honorable Max Baucus
Chairman

The Honorable David Vitter
Ranking Member
Subcommittee on Transportation and Infrastructure
Committee on Environment and Public Works
United States Senate

The Transportation Infrastructure Finance and Innovation Act (TIFIA) program provides federal credit assistance in the form of direct loans, loan guarantees, and lines of credit to finance surface transportation projects including highway, transit, rail, port access, and intermodal projects. Created in 1998 as part of the Transportation Equity Act for the 21st Century (TEA-21),[1] the program is designed to fill market gaps and leverage substantial private and other nonfederal investment to help advance projects of regional and national significance. For most of its history, the TIFIA program was underutilized. Beginning in fiscal year 2008, however, interest in the program increased substantially because of several factors, including the growing demand for infrastructure investment relative to available transportation funding and the economic downturn. For the past 3 years, demand for credit assistance has exceeded the program's budget resources by a factor of more than 10 to 1. In part because of to this oversubscription, Members of Congress and others have offered reauthorization proposals to greatly expand and modify various aspects of the program.

The Federal Highway Administration (FHWA) within the U.S. Department of Transportation (DOT) administers the TIFIA program. DOT awards credit assistance to eligible project sponsors, such as state departments of transportation, transit agencies, tolling authorities, and private entities. To receive assistance, projects must meet several eligibility requirements—such as having a total cost of at least $50 million[2]—and sponsors must submit a letter of interest (LOI) and a formal application. Since the program's inception, 26 projects in 12 states, the District of Columbia, and Puerto Rico have received about $9.1 billion in TIFIA credit assistance through direct loans and one loan guarantee, and over a dozen more projects are working toward receiving credit assistance.

In response to your request, this report addresses the following questions: (1) What are the characteristics of TIFIA projects and how does DOT track progress toward the program's stated goals? (2) How did DOT evaluate and select projects to apply for TIFIA credit assistance for fiscal years 2010 and 2011? (3) What factors affect states' and other project sponsors' decisions about whether to apply for TIFIA credit assistance? (4) What are the proposed options to modify the TIFIA program?

To address these objectives, we reviewed DOT program guidance for TIFIA, relevant legislation and regulations, and DOT's biennial reports to Congress on the TIFIA program. We also reviewed documents developed by DOT to capture the evaluation and scoring decisions for projects seeking credit assistance in fiscal years 2010 and 2011. We analyzed data from DOT on LOIs submitted to the program since its creation and on projects in the TIFIA portfolio—projects with credit agreements—to describe the characteristics and results of the TIFIA program. We assessed the reliability of the data by reviewing DOT's data documentation, interviewing knowledgeable officials, and conducting independent validation through use of our web survey of states and recent applicants. We found the data to be sufficiently reliable for our purposes. In addition, we interviewed DOT officials on TIFIA program goals and performance measures, the selection and evaluation of projects, the demand for the program, and proposed options to modify TIFIA. We also interviewed project sponsors, including those that received credit assistance in California, Colorado, Florida, Texas, and Virginia, the 5 states with the most TIFIA awards; the state DOTs in North Carolina and Iowa; private concessionaires, private sector entities that have entered into agreements with public agencies to allow greater private sector participation in the delivery and financing of transportation projects; and legal and financial advisors that have been involved in securing TIFIA credit assistance to learn about their experience with the TIFIA program, including the application process, oversight and monitoring of credit assistance, future demand for the program, and proposed options to change the program.

In addition, we surveyed all state departments of transportation and all recent applicants that submitted an LOI to the TIFIA program in fiscal years 2010 and 2011 using a web-based survey.[3] The survey recipients fell into four categories: (1) state DOTs from states from which no sponsor had ever applied to the TIFIA program, (2) state DOTs from states for which a sponsor had applied to the TIFIA program but not in recent years—that is, 2010 or 2011, (3) state DOTs who had recently applied to the TIFIA program, and (4) other non-state DOT organizations, such as transit agencies, who had recently applied to the TIFIA program. Survey respondents were presented with different questions in the survey depending on their past experience with the TIFIA program and whether they were state DOTs. In general, the survey topics included the factors contributing to organizations' decisions about whether to seek TIFIA assistance, satisfaction with the process for submitting an LOI to the TIFIA program, opinions on proposed modifications to the TIFIA program, potential future demand for the TIFIA program, and characteristics of the state DOTs. We distributed a link for the survey to 83 organizations by email and also subsequently e-mailed and telephoned nonrespondents to encourage a higher response rate. We received completed surveys from 66 respondents (80 percent). Of the 66 completed responses, 16 were state DOTs from states with no TIFIA experience, 14 were from state DOTs from states with no recent TIFIA experience, 12 were from state DOTs that were recent applicants, and 24 were from non-state DOT organizations that were recent applicants.

To determine options that had been proposed to change the TIFIA program, we reviewed the American Energy and Infrastructure Jobs Act of 2012,[4] as reported out by the House Committee on Transportation and Infrastructure, and the Moving Ahead for Progress in the 21st Century Act (MAP-21),[5] as passed by the Senate. We also reviewed other reauthorization proposals and documents prepared by, among others, DOT and the American Association of State Highway and Transportation Officials (AASHTO). In interviews, we discussed these and other options with DOT, select project sponsors, legal and financial advisors, and private concessionaires with investments in transportation infrastructure, all familiar with or recipients of TIFIA assistance.

We conducted this performance audit from July 2011 to June 2012 in accordance with generally accepted government auditing standards. Those standards require that we plan and perform the audit to obtain sufficient, appropriate evidence to provide a reasonable basis for our findings and conclusions based on our audit objectives. We believe that the evidence obtained provides a reasonable basis for our findings and conclusions based on our audit objectives. For more information on our scope and methodology, see appendix I.

BACKGROUND

Most federal highway transportation funds are distributed as grants to states as part of the federal aid highway program through a set of complex formulas that take into account a number of factors, including the estimated share of taxes highway users in each state contributes. The Highway Trust Fund is the principal source of funding for federal aid highway programs and is funded through motor fuel and other highway use taxes.[6] Grants for transit projects are distributed as part of the federal transit program through a collection of formula-based and discretionary programs and are funded primarily by the Mass Transit Account of the Highway Trust Fund.[7] Supplementing these federal programs is a collection of financing methods that allow project sponsors—such as state DOTs and transit agencies—to borrow money through bonds, loans, or other mechanisms. State DOTs and other project sponsors can raise money in the bond market through, for example, revenue bonds backed by anticipated project revenues like tolls;

bonds backed by future federal transportation funds, such as Grant Anticipation Revenue Vehicles (GARVEE) or Grant Anticipation Notes (GAN); or general obligation bonds backed by the full faith and credit of a state or municipality. Project sponsors may also seek private investment through bank debt, private equity, or private activity bonds. Through TIFIA, DOT provides loans or other credit assistance to sponsors of surface transportation projects. Declining Highway Trust Fund revenues and states' budget constraints, as well as the high cost and size of major transportation projects, have prompted project sponsors to seek alternative methods of funding transportation infrastructure.

TIFIA Goals

The TIFIA program's primary goal is to leverage limited federal resources and stimulate private capital investment in transportation infrastructure by providing credit assistance to projects of national or regional significance. Underlying the TIFIA program is the notion that the federal government can perform a constructive role in financing large transportation infrastructure projects by supplementing, but not supplanting, existing capital finance markets. In this role, DOT identifies five key objectives for the TIFIA program:

- facilitate projects with significant public benefits;
- encourage new revenue streams and private participation;
- fill capital market gaps for secondary (subordinate) capital;
- be a flexible, "patient" investor willing to take on investor concerns about investment horizon, liquidity, predictability, and risk; and
- limit federal exposure by relying on market discipline.

TIFIA Credit Assistance Terms and Eligibility Requirements

DOT provides TIFIA credit assistance in three forms: direct loans, loan guarantees, and standby lines of credit. The maximum maturity for all types of TIFIA credit assistance is 35 years after substantial completion of a project. Lines of credit can supplement project revenues during the first 10 years of project operations. In addition, DOT can defer the first TIFIA repayment until

5 years after substantial completion of a project, and most project sponsors avail themselves of this option.

Other credit assistance terms include that (1) TIFIA assistance may provide no more than 33 percent of total project costs,[8] (2) senior debt has an investment-grade credit rating (Baa3/BBB- or higher),[9] and (3) TIFIA assistance can be subordinate to the project's senior debt, meaning that senior creditors may receive project revenues ahead of DOT.

According to DOT officials, the TIFIA program is one of the few federal credit programs in which federal assistance routinely takes a subordinate position to other, nonfederal lenders with respect to cash flows. However, to protect taxpayers, TIFIA loans may not be subordinated to the claims of other creditors with respect to the loan recipients' bankruptcy, insolvency, or liquidation.[10]

Both public and private entities are eligible to receive TIFIA assistance for a range of surface transportation-related projects including highway, transit, railroad, intermodal freight, and port access. Borrowers can include entities like state DOTs, toll authorities, transit agencies, and private concessionaires.[11] Other eligibility requirements include that a project

- must have total costs of at least $50 million (or $15 million for intelligent transportation systems projects); 12
- must be included in state and local transportation plans; and
- must have dedicated revenues, like tolls, user fees, or pledged taxes, for repayment.

Through the Safe, Accountable, Flexible, Efficient Transportation Equity Act: A Legacy for Users (SAFETEA-LU), Congress amended and reauthorized the TIFIA program, authorizing budget authority of $122 million for each of fiscal years 2005-2009 from the Highway Trust Fund for the program's credit subsidy cost and administrative expenses.[13] The credit subsidy is the estimated long-term cost to the government of providing assistance.[14]

Extensions of SAFETEA-LU have authorized budget authority of $122 million for the TIFIA program for each subsequent fiscal year. Any uncommitted budget authority remains available for obligation in subsequent years, unless Congress chooses to reprogram or rescind these amounts. According to DOT, $10 million in TIFIA budget authority can generally be leveraged to provide $100 million in credit assistance.[15]

TIFIA Program History

In fiscal year 2008, total requests for TIFIA assistance exceeded DOT's available budgetary resources for the first time. Prior to this, when there was lower demand for the program, DOT allowed project sponsors to seek TIFIA assistance on a "first come, first served" basis defined by the sponsor's schedule.[16]

Figure 1 shows the number and amount of credit assistance requested each fiscal year.

In its 2010 report to Congress on TIFIA, DOT attributed the increased demand for TIFIA assistance to several factors, including the growing demand for infrastructure investment relative to other sources of funding (like declining fuel tax receipts), the economic downturn and difficulty accessing capital markets, and the increasing use of innovative approaches, like public-private partnerships, to finance and deliver projects.

After demand exceeded available budget resources, DOT terminated the open application process and instituted an annual, fixed-date solicitation process for sponsors to submit LOIs for credit assistance. In fiscal year 2010, DOT began evaluating and competitively selecting projects based on how well they align with the TIFIA selection criteria and the availability of budget resources.[17]

Primary Stages for Securing a TIFIA Credit Agreement

As shown in figure 2, there are four primary stages for securing TIFIA credit assistance. For each project, the amount of time needed to complete each stage varies. For instance, in its 2002 report to Congress, DOT stated that the length of credit agreement negotiations is affected by the complexities and uncertainties of large infrastructure projects as well as the learning curve of both project sponsors and DOT as they encounter unique legal and financial issues with projects.

After determining that a project meets the eligibility requirements, DOT uses eight statutory criteria[18] weighted by regulation[19] to select projects to receive credit assistance.

Table 1. DOT Definitions and Clarifications of TIFIA Statutory Criteria

Criterion (weight)	Definition	2010 and 2011 clarifications
National or regional significance (20%)	Extent to which project is nationally or regionally significant in terms of generating economic benefits, supporting international commerce, or otherwise enhancing national transportation system. It includes • livability—providing transportation options linked with housing and commercial development to improve the economic opportunities and quality of life for people in communities across the United States, • economic competitiveness—contributing to the economic competitiveness of the United States by improving long-term efficiency and reliability in the movement of people and goods, and • safety—improving the safety of U.S. transportation facilities and systems and the communities and populations they affect.	Livability Economic competitiveness Safety
Environment (20%)	Extent to which the project helps maintain or protect the environment. It includes • sustainability—improving energy efficiency, reducing dependence on oil, reducing greenhouse gas emissions and other transportation-related impacts on ecosystems through use of tolling or pricing structures to reduce congestion and encourage the use of alternative transportation options, and • state of good repair—improving the condition of existing transportation infrastructure with particular emphasis on projects that minimize life cycle costs and use environmentally sustainable practices and materials.	Sustainability State of good repair
Private participation (20%)	and attract private debt or equity investment.	
Project acceleration(12.5 %)	Likelihood that TIFIA assistance would enable a project to proceed at an earlier date than the project would otherwise be able to proceed.	
Creditworthiness (12.5%)	The creditworthiness of the project, including a determination by the Secretary of Transportation that any financing for the project has appropriate security features to ensure repayment, such as rate covenants.	

Table 1. (Continued)

Criterion (weight)	Definition	2010 and 2011 clarifications
Use of new technologies (5%)	Extent to which project uses new technologies, including intelligent transportation systems, to enhance project efficiency.	
Reduced federal grant assistance (5%)	Extent to which TIFIA assistance would reduce the contribution of federal grant assistance to the project.	
Consumption of budget authority (5%)	Amount of budget authority required to fund the federal credit instrument made available under TIFIA.	

Source: GAO analysis of DOT documents.

Note: On the basis of both the funding notices for fiscal years 2010 and 2011, DOT did not consider two criteria—creditworthiness and consumption of budget authority—when evaluating LOIs as DOT did not believe project sponsors would have sufficient details regarding these criteria at the time of the LOI submission. Per the funding notice issued on November 3, 2011, DOT considered all eight criteria in its evaluation of LOIs beginning in fiscal year 2012.

Beginning in fiscal year 2010, DOT defined the statutory criteria in the notice of funding availability (funding notice), which included clarification of the national or regional significance and environment criteria. (See table 1.)

In addition, projects that received funding through DOT's Transportation Investment Generating Economic Recovery (TIGER) grant program were eligible for TIFIA credit assistance. Specifically, project sponsors could use TIGER grant funds for TIFIA credit assistance, known as a TIGER TIFIA payment. In each of the four rounds through which DOT TIGER grants have been available, a portion of the funds could be used to support the credit subsidy and administrative costs of projects eligible for federal credit assistance. When a project sponsor is offered a TIFIA award through the TIGER program, the project goes through the regular TIFIA process, including the TIFIA credit evaluation, credit agreement negotiation, and oversight and monitoring.[20]

DOT implements and manages the program using internal staff and a pool of external financial and legal advisors. There are nine internal TIFIA office staff who are responsible for assisting with reviewing LOIs, selecting projects to apply for credit assistance, negotiating credit agreements, monitoring loan disbursements and the financial performance of executed credit agreements, and tracking credit subsidy calculations.

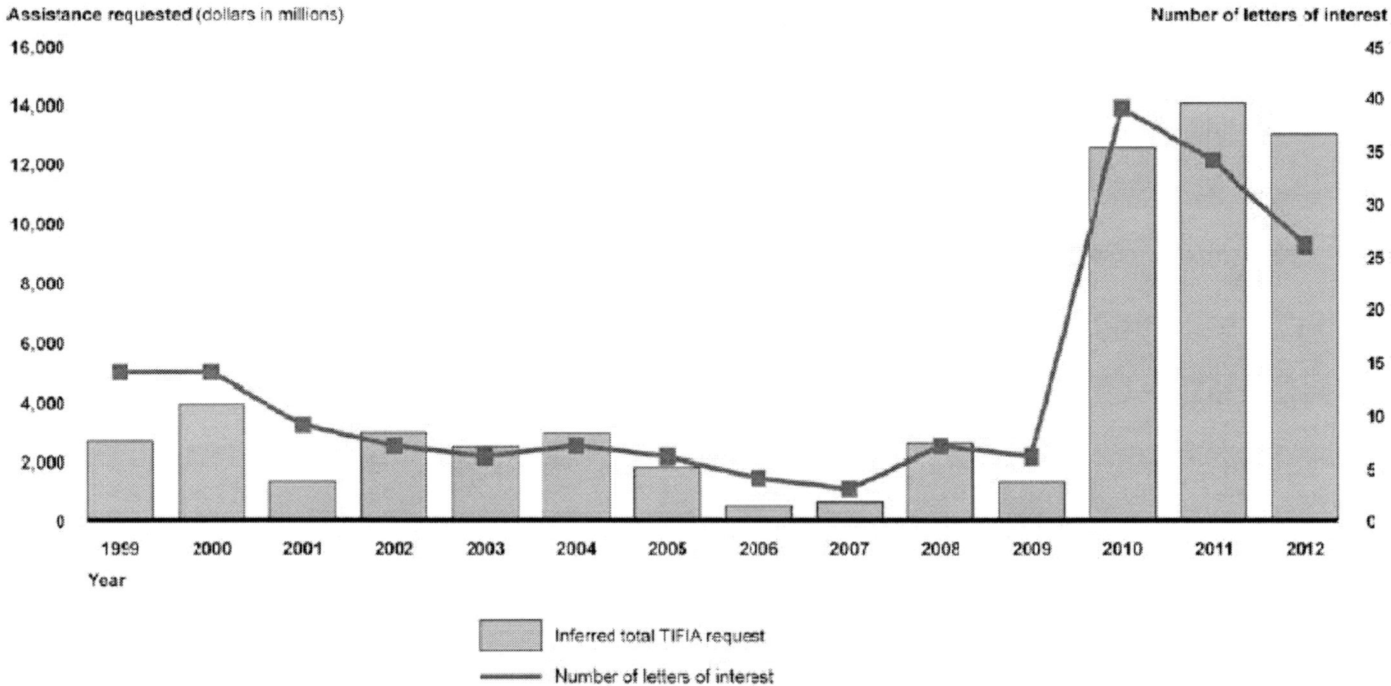

Source: GAO analysis of DOT data on letters of interest.

Figure 1. Requests for TIFIA Credit Assistance, Fiscal Years 1999-2012.

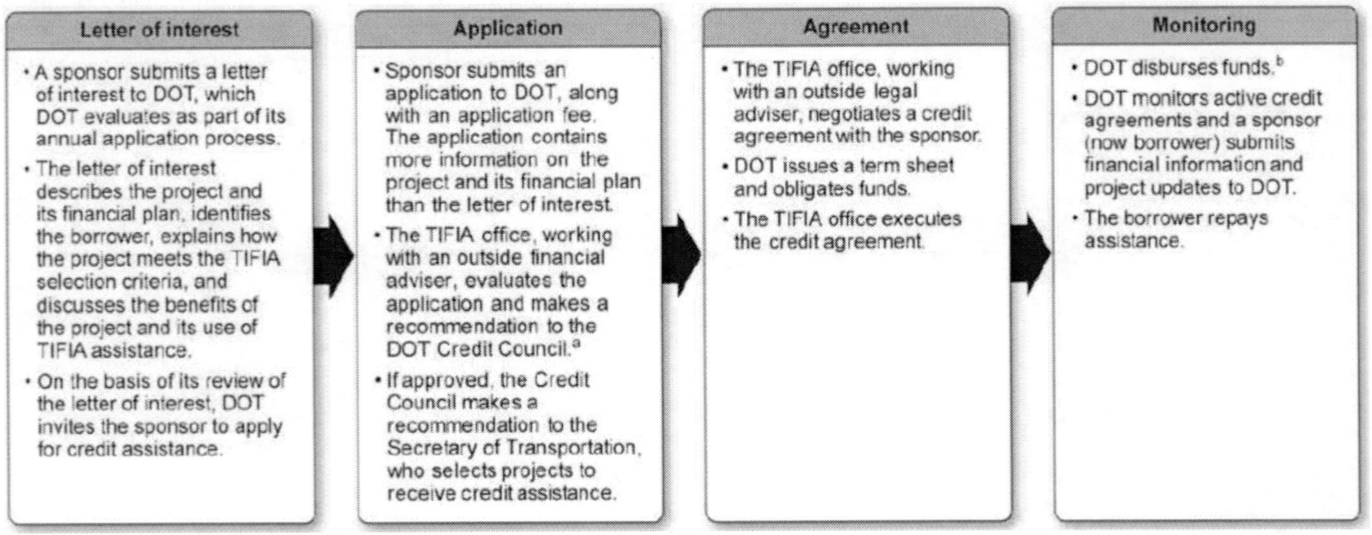

Source: GAO analysis of DOT documents.

Notes: Sponsors of highway projects may seek waivers from the usual TIFIA application process through FHWA's Special Experimental Project Number 15 (SEP-15). SEP-15 allows for the modification of FHWA policy and procedure to facilitate public-private partnerships and other types of innovation in the federal aid highway process. Specifically, sponsors can seek waivers from FHWA procedures and policy, including those that apply to TIFIA.

[a]The Credit Council provides policy direction to the TIFIA program and recommends projects to receive credit assistance. The DOT Credit Council is composed of the following DOT officials: the Deputy Secretary of Transportation, who serves as chair; Assistant Secretary for Budget and Programs, who serves as vice-chair; Under Secretary of Transportation for Policy; General Counsel; Assistant Secretary for Transportation Policy; and the Director of the Office of Small and Disadvantaged Business Utilization. The

Federal Highway Administrator, Federal Transit Administrator, Federal Railroad Administrator, and Maritime Administrator also sit on the Credit Council. Finally, there can be up to three at-large members.

[b]For loan guarantees, DOT disburses funds to the guaranteed lender only when there is a payment default by the borrower. Any funds that the guaranteed lender draws from the loan guarantee initiate a loan between DOT and the borrower.

Figure 2. Four Stages of TIFIA Credit Assistance.

Currently, DOT supplements its internal staff with a pool of five financial and legal advisors, contracted for 5-year periods, who assist in the review of applications and negotiation of credit agreements.[21]

TIFIA PROJECTS ARE MOSTLY LARGE HIGHWAY PROJECTS; PERFORMANCE MEASURES ARE NEEDED TO BETTER EVALUATE PROGRAM OUTCOMES

TIFIA Projects Are Concentrated in Five States and Tend to Be Large Highway Projects

Since the TIFIA program was created in 1998, DOT has executed 27 credit agreements for 26 projects.[22] To date, assistance has been provided through 26 loans and one loan guarantee. Of the 26 projects, 17 are located in 5 states—California, Colorado, Florida, Texas, and Virginia (see fig. 3).

Overall, sponsors from 33 states, the District of Columbia, and Puerto Rico have submitted LOIs for projects that vary by mode and purpose, but most have high total costs. Highway projects account for a majority of all LOIs submitted to TIFIA program.

According to DOT data, highway projects—such as building new roads and replacing bridges—accounted for about 60 percent of the 182 LOIs submitted to the TIFIA program from 1999 to 2012. Transit and intermodal projects—such as building new transit systems and constructing parking garages and facilities linking various transport modes near airports—account for 18 percent and 10 percent of all LOIs, respectively. In addition, rail, ferry, and ITS projects account for 4 percent, 2 percent, and 1 percent of LOIs during this time, respectively.[23]

However, no projects in these three modes have received TIFIA assistance to date. Over the history of the program, the average total cost of projects seeking TIFIA assistance has been $1.2 billion. Through fiscal year 2012, no sponsor in the other 17 states has submitted an LOI to the TIFIA program.24 (See fig. 4.)

According to DOT data, TIFIA credit agreements have been used mostly for large, high-cost highway projects. Overall, DOT has provided TIFIA assistance to 17 highway projects.

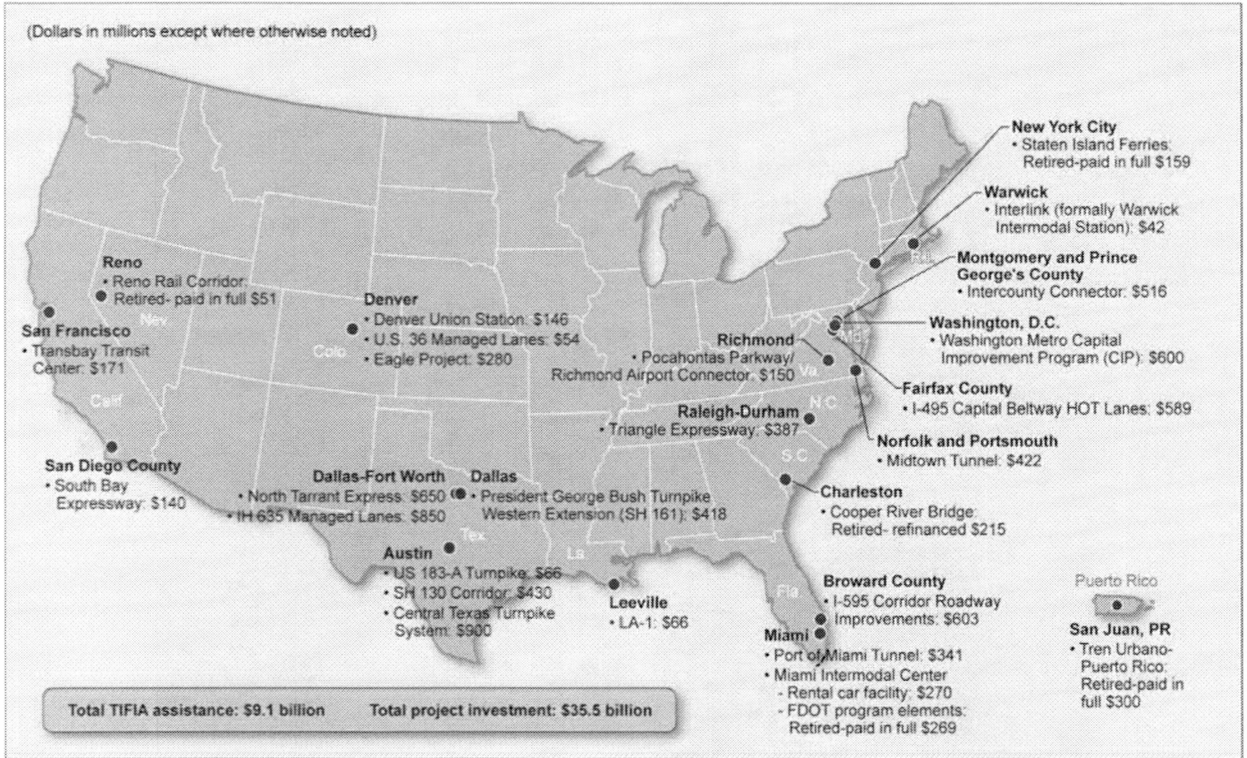

Sources: DOT and Map Recources.

Note: Dollar values indicate the amount of TIFIA assistance each project received, in millions of dollars.

Figure 3. Location of Projects with TIFIA Credit Agreements.

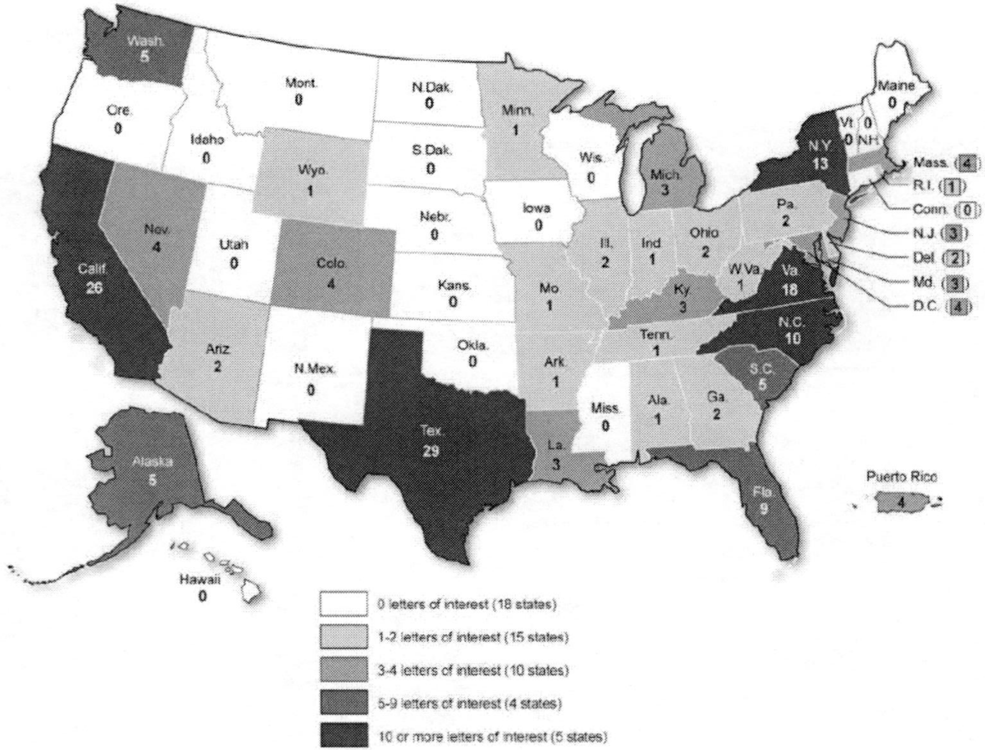

Sources: GAO analysis of DOT information and Map Recources.

Note: Counts include duplicate LOIs submitted for the same project in different years.

Figure 4. States That Had Submitted TIFIA Letters of Interest, through Fiscal Year 2012.

Some of these projects—like the President George Bush Turnpike-Western Extension (SH 161) in Texas— were to construct new roads, and others—like the I-595 Corridor Roadway Improvements project in Florida—to reconstruct and expand existing roads. Projects receiving credit assistance also tend to have high total costs. Of the 25 projects, 20 projects cost more than $500 million and 16 projects cost more than $1 billion. The average total cost of projects receiving TIFIA credit assistance is $1.4 billion.[25]

According to DOT, TIFIA assistance can help advance large-scale projects that otherwise might be delayed or deferred because of size or complexity, and as such, TIFIA projects to date have mainly been large-scale projects. On average, TIFIA assistance accounts for 24 percent of total project costs, about 9 percent less than the 33 percent currently permitted by law.

To a lesser extent, TIFIA has also been used for transit and intermodal projects. Four intermodal projects have received credit assistance, including the Reno ReTRAC project in Nevada, which includes rail and roadway improvements to improve freight capacity and address environmental and safety concerns. Five transit projects have received TIFIA assistance. The Tren Urbano project in Puerto Rico, for example, constructed a new, fixed-guideway transit system to relieve congestion in the San Juan area. DOT officials told us that the balance of projects is becoming more diverse in terms of mode. They noted that sponsors of transit projects have been slower to use TIFIA assistance in the past, primarily because transit projects have access to low-cost municipal debt and do not generate revenue in excess of their operating costs to repay assistance. Moreover, it can be difficult to integrate TIFIA assistance with federal funding for transit provided through the New Starts program. [26] However, in fiscal years 2010 and 2011, DOT invited the sponsors of 12 projects to apply for TIFIA assistance, which is the next stage in securing TIFIA assistance, 4 of which were transit projects.[27]

Projects financed with TIFIA credit agreements also have other attributes. According to DOT data, roughly one-third of projects are public-private partnerships that include private equity investments. For these projects, the private equity accounts for about 17 percent of total project costs. Defining private participation more broadly, 17 projects with active credit agreements include either private equity or debt. The average private investment for projects with active credit agreements, including equity and debt, is 37 percent of total project costs.[28] The North Tarrant Express, for example, is a public-private partnership between the Texas Department of Transportation and a private concessionaire—NTE Mobility Partners— to design, build, finance, operate, and maintain a 13-mile section of highway in the Dallas-Fort Worth area. Project funds include $426 million in equity from NTE Mobility Partners and $398 million from private activity bonds. These two sources of private investment account for about 40 percent of the project's total cost. Projects with credit agreements typically pledge user fees or dedicated tax revenue to repay TIFIA assistance. For 16 credit agreements, user fees like tolls are pledged to repay assistance, while for 8 credit agreements, tax-backed revenue streams like local sales taxes are pledged to repay assistance. The remaining 3 credit agreements use other dedicated revenues, like availability payments,[29] to repay assistance.[30]

As of April 2012, DOT reported that it has provided nearly $9.1 billion—$8.5 billion through direct loans and $600 million in loan guarantees—to projects at a budgetary cost of about $654 million. The budgetary cost of TIFIA assistance is the total credit subsidy for all projects, with the credit subsidy, as noted earlier, being the estimated long-term cost to the government of providing assistance calculated on a net present value basis, excluding administrative costs. As such, the credit subsidy reflects the estimated risk of the loan or assistance. According to DOT, the original credit subsidy cost for credit agreements ranges from less than 1 percent to over 15 percent of the amount of TIFIA assistance. Projects that pledge user fees tend to have higher subsidy costs and, thus, generally entail greater risk to the federal government because actual usage and fees for a project (such as traffic and toll revenue on a new road) may not meet projections, particularly early in its operation. In such cases, where repayment of TIFIA assistance relies solely on revenues from user fees, poor performance—such as less than projected use of a facility—could result in nonpayment.

Most Sponsors Are Drawing Funds for Projects

Project sponsors are actively drawing funds from DOT for about half of the projects with TIFIA credit agreements.[31] Many of these projects—14 of 26 projects—are currently under construction. As a result, many sponsors are drawing and not yet repaying TIFIA loans. Six project sponsors have retired their TIFIA credit agreements through early repayment, by refinancing the loan, or because of expiration of the credit agreement in the case of a loan guarantee. For example, the Puerto Rico Highway and Transportation Authority refinanced its TIFIA loan for the Tren Urbano project with tax-exempt debt about 3 years after DOT fully disbursed the loan. The sponsor paid back the TIFIA loan 32 years ahead of schedule and anticipated saving about $31.7 million in interest payments to DOT by refinancing the TIFIA loan.

The sponsor of one TIFIA project—the South Bay Expressway in San Diego County, California—declared bankruptcy in 2010 but has not defaulted on any TIFIA payments. At the time of the bankruptcy filing, the outstanding balance of the TIFIA loan was $172 million, including interest. The Plan of Reorganization ordered by the U.S. bankruptcy court reduced the value of the loan's principal. DOT's unsecured claim was $73 million, or 42 percent of the outstanding loan balance. Following the sale of the project to and the assumption of the TIFIA loan by the San Diego Association of Governments (SANDAG) in December 2011, DOT expects to recover the original loan value through higher interest rates charged on the restructured loan.

TIFIA Program Could Benefit from Use of Performance Measures to Evaluate Progress toward Goals

While DOT tracks certain aspects of a project, such as monitoring whether the project is meeting its construction timeline and comparing actual and projected receipts of the revenue pledged to repay TIFIA assistance, the agency does not systematically assess whether its portfolio as a whole is achieving the program's goals of leveraging federal funds and encouraging private co-investment. The Secretary of Transportation is required to report biennially to Congress on the financial performance of projects that are receiving or have received TIFIA assistance and whether the goals of the TIFIA program are best served by continuing the program under the authority of the Secretary, establishing another entity to administer to program, or phasing out the program.[32]

As DOT administers a growing portfolio of TIFIA credit agreements, decision making and program implementation could be aided by additional use of performance measures to assess how well the agency is meeting its goals. Moreover, knowing the degree to which the agency is meeting its goals could augment the information DOT uses to fulfill the above stated requirement to make recommendations to Congress. According to our work on leading performance management practices, agencies should have plans for managing their programs that identify goals, strategies, time frames, resources, and stakeholder involvement in decision making.[33] Though DOT has identified goals and objectives for the TIFIA program, its limited use of performance measures makes it difficult to determine the degree to which the TIFIA program is meeting these goals and objectives. Other programs administered by DOT do have performance measures.

Programs administered by the Federal Railroad Administration (FRA) and the Federal Motor Carrier Safety Administration, for example, have created a set of performance goals and measures to address program performance.[34] For example, FRA set goals to reduce the rate of train accidents in its proposed fiscal year 2013 budget, and FRA tracks these goals and actual accident rates over time to measure whether or not it is meeting its safety goals. Also, for FHWA's Express Lanes Demonstration program, DOT developed performance measures to evaluate projects' performance along four program goals— such as travel, traffic, and air quality—and uses information collected annually from project sponsors to report to Congress on the projects' performance.[35]

In its first report to Congress in 2002, DOT examined the extent to which projects approved to receive assistance collectively met key TIFIA goals and objectives.

For instance, DOT calculated that the amount of private co-investment in projects totaled $3.1 billion, or about 20 percent of the projects' total costs.[36] DOT also calculated that TIFIA had a federal leverage ratio of 4.8, meaning that every dollar of federal investment in projects approved to receive assistance—including TIFIA as well as other federal funds—represented nearly $5 in total infrastructure investment. However, DOT has not presented similar data on its progress in meeting the program's goals in any of the subsequent reports to Congress.

In these reports, DOT provides only broad descriptive information on the financial status of the projects and highlights project innovations, such as reporting on the use of new revenue sources like availability payments to repay assistance. DOT provides information on each project with a TIFIA credit agreement that describes each project and lists funding sources, but DOT does not aggregate this information for the portfolio of projects on its website or in other program documents. Given that the agency collects such data on projects that received TIFIA assistance, it could use these data—in particular, the amount of federal and nonfederal funding and financing, as well as the amount of private equity and debt—to better evaluate the progress toward meeting program goals and objectives, like leveraging limited federal resources and stimulating private capital investment in transportation infrastructure.

DOT USED A COMPETITIVE PROCESS TO EVALUATE PROJECTS, BUT THIS PROCESS COULD BENEFIT FROM INCREASED TRANSPARENCY

DOT Developed a Competitive, Two-Step Process to Evaluate and Select Projects Using the Statutory Criteria

In response to increased program demand and the uncertain budget environment, DOT used a competitive, two-step process to assess LOIs and select projects to apply for credit assistance in fiscal years 2010 and 2011.[37] DOT officials told us that they began using the current evaluation process—focusing on the LOIs to pre-assess a project's alignment with TIFIA's statutory criteria—to address the significant increase in demand for the program coupled with the current uncertain and limited budgetary environment due in part to a lack of a long-term surface transportation reauthorization bill. These circumstances, according to DOT, required the agency to establish a process that allows the agency to choose amongst best-qualified projects in each fiscal year instead of accepting eligible projects on a first-come, first-served basis as was the case when the program was undersubscribed.

DOT's process for competitively selecting amongst LOIs involved two steps; first, DOT convened a multimodal team to assess, score, and group projects using statutory criteria, and second, DOT used a team of senior-level staff—called the executive leadership team—to review the multimodal team's assessments and invite select project sponsors to submit an application for credit assistance (see fig. 5).

Multimodal Team

The multimodal team—composed of staff from different DOT modal administrations[38]—individually assessed each LOI against the statutory criteria to assign preliminary scores. Multimodal team members read and assigned each LOI a numeric score of 0 to 4 for each of the six criteria, with 0 indicating that a project was not consistent with a criterion and 4 indicating that a project was most consistent with a statutory criterion.[39] While the funding notices defined each of the statutory criteria, they did not describe specific project qualifications or benefits that would merit a higher or lower score.

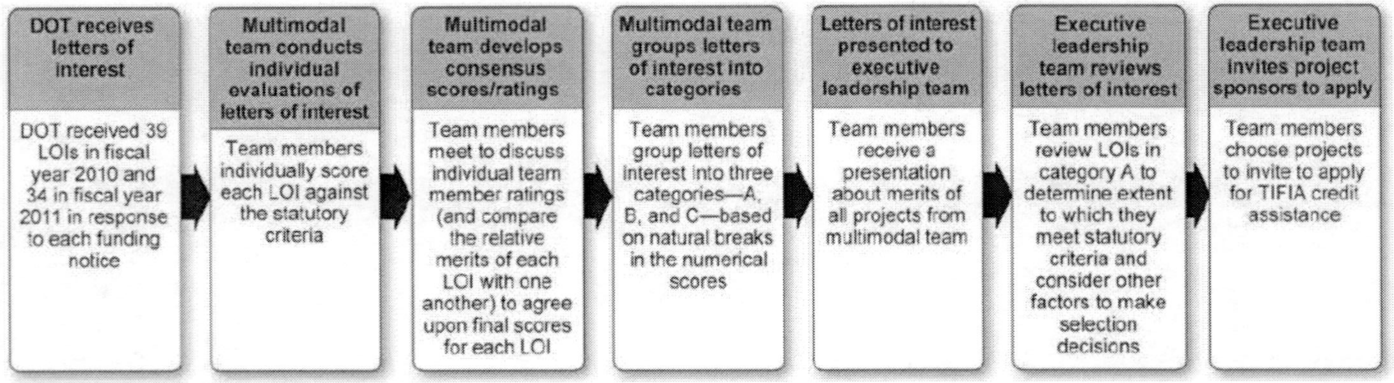

Source: GAO analysis of DOT documents.

Figure 5. TIFIA LOI Evaluation and Selection Process for Fiscal Years 2010 and 2011.

Additionally, multimodal team members did not use any guidance beyond the funding notices to delineate what the possible range of scores signified in terms of project qualifications and benefits. DOT officials said that evaluators relied primarily on content in the submitted LOIs, as well as their own modal expertise as necessary, to evaluate projects. Each LOI contained information to describe the project and its proposed financial plan, identify the proposed borrower, explain how the TIFIA statutory selection criteria are met, and describe the benefits of the proposed project and its use of TIFIA assistance.

To finalize their individual scores for each LOI, multimodal team members compared LOIs with one another to determine the relative merits of each project when assigning scores. For example, if an LOI received a preliminary score of 3.5 for the private participation criterion, but when compared with other projects in its cohort appeared less well aligned with the private participation criterion, its score would be lowered to reflect its relative rank among the LOIs.

Also, multimodal team members met over several weeks to discuss and compare LOIs in an effort to help ensure reliability in scoring across team members.

To arrive at a final score for each LOI, individual team members' final scores were combined for each criterion and a cumulative weighted total score based on assigned weights in regulation was calculated for each project.

Last, the multimodal team rank ordered the LOIs by the total score and grouped them into three categories—A, B, and C. DOT officials said that the multimodal team grouped LOIs based on natural breaks in the numerical scores. LOIs placed in category A were those that scored the highest numerically, and thus were considered to be the most consistent with the statutory criteria. Table 2 details the category grouping for LOIs by fiscal year.

Table 2. Number of LOIs, by Fiscal Year and Outcome of Multimodal Team's Assessment

Fiscal year	Number of LOIs in category A	Number of LOIs in category B	Number of LOIs in category C	Total
2010	7	17	15	39
2011	21	6	5	32[a]

Source: GAO analysis of DOT data.

[a] DOT did not assign a category for two additional LOIs in 2011, as these projects had received TIGER grants to cover the subsidy cost of a TIFIA loan.

As shown in table 2, there was a threefold increase in the number of category A LOIs from fiscal year 2010 to fiscal year 2011. DOT officials attributed this increase to higher-quality LOI submissions as well as improved project readiness of resubmitted projects.

For projects that submitted an LOI in fiscal year 2011, 10 of the 34 had previously submitted an LOI in 2010.

For projects that submitted an LOI in fiscal year 2012, 16 of 26 had applied in either fiscal year 2010 or 2011.

Executive Leadership Team

After the multimodal team grouped LOIs and provided a briefing about its assessment of all projects to the executive leadership team,[40] this second team reviewed the projects and selected a subset of the category A projects to advance.

According to DOT officials, in its review of projects, the executive leadership team was not aware of the scoring or ranking distinctions amongst LOIs in the category because numerical scores assigned by the multimodal team were removed. Instead, only basic project information, including high-level project summaries and category groupings (A, B, or C), were provided to the executive leadership team.[41] Similar to the multimodal team evaluation, the executive leadership team did not use any guidance beyond the funding notice in its review of LOIs and relied primarily on content in the LOIs to score projects. However, in some cases, the team sought clarification from DOT staff, including FHWA division offices in various states, to gather additional information on a project's readiness, like its status in the environmental review process.

The executive leadership team reviewed the LOIs to determine the extent to which they met statutory criteria and, to choose projects to advance from among category A LOIs, considered other factors including available budget authority and geographic diversity.[42] In making final selections, the executive leadership team had to consider projects that could be funded within TIFIA's limited budget authority. Further, with limited budget authority, DOT officials said that project acceleration—which encompasses factors like the project's progress in completing environmental review requirements—was an important consideration in picking among projects that were consistent with the statutory criteria, particularly in 2011, when there were a higher number of category A LOIs.

Table 3. Number of LOIs Invited to Apply, by Fiscal Year and Mode

Fiscal year	Total number of projects	Highway	Transit	Intermodal	Other
2010					
Invited to apply	4	2	1	0	1
Total	39	29	5	3	2
2011					
Invited to apply	8	4	3	1	0
Total	34	24	8	1	1

Source: GAO analysis of DOT data.

Note: Other includes projects like those for truck technology systems and multimodal transportation improvements.

Table 4. Status of Projects Invited to Apply, by Fiscal Year

Fiscal year	Total number invited to submit a full application	Number with an executed credit agreement	Number with an application under review	Number for which DOT is awaiting an application
2010	4	1	2	1
2011	8	1	2	5

Source: DOT, as of April 2012.

As a result of its review, the executive leadership team invited 4 projects in fiscal year 2010 and 8 projects in fiscal year 2011 to submit a full TIFIA application.43 (See table 3.)

Overall, this two-step process ensured that projects invited to apply were from among the highest-scoring LOIs overall—that is, from category A— but did not ensure that the projects selected were those that scored the highest numerically by the multimodal team. According to DOT officials, relying on numerical scores alone could provide a false sense of precision in selecting projects to advance. To date, because only category A projects are forwarded, no category B or C projects from the multimodal team evaluation have advanced over category A projects in the executive leadership team evaluation. DOT officials said that while there are no specific requirements to do so, the executive leadership team has only considered advancing LOIs in category A.

After the executive leadership team has invited the sponsors of projects to apply for credit assistance, project sponsors must submit a full TIFIA application, after which DOT conducts a full evaluation of the application and makes a recommendation to the Credit Council. Then, the Secretary of

Transportation makes the final decision on whether to approve a project to receive TIFIA credit assistance. Six of the 12 of the project sponsors that were invited to apply in fiscal years 2010 and 2011 have not yet submitted an application to DOT but all are still pursuing TIFIA loans. (See table 4.)

DOT officials and project sponsors that had executed TIFIA credit agreements said that the amount of time it takes for sponsors to complete the application and negotiation process varies by project. Several factors—such as the status of a project's environmental review, the complexity of the project's finance and delivery approach, and changes to the project—can influence the length of these processes. According to DOT, sponsors of four invited projects that have not yet submitted an application are completing work to comply with federal environmental requirements. Also, some project sponsors we spoke with said that the TIFIA application and negotiation processes can be longer for projects that have more complex financial plans, such as having a less frequently used revenue stream or relying on future state appropriations. For example, after being invited to apply, one sponsor we interviewed had to complete the process to select a private concessionaire for the project; then, since submitting the TIFIA application, the sponsor has been working with the TIFIA office regarding uncertainty around appropriations from the state legislature before beginning the negotiation process.

For projects not invited to apply, staff from the TIFIA office provided feedback to sponsors on their LOIs upon request. According to DOT officials, the primary aim of feedback is to explain how the LOI performed against each criterion. In addition, feedback included information on how a project sponsor could improve an LOI for resubmission, such as explaining that it needs to provide more details on specific project benefits. However, through this feedback, a project sponsor is not informed about the numeric scoring or ranking of its LOI relative to other LOIs. In some cases, DOT officials said that the feedback provided indicated that there was nothing "wrong" with a project's LOI but that it was not invited to apply given the strength of the pool of LOIs submitted in that round.

TIFIA Selection Process Could Benefit from Greater Transparency

While project sponsors and other stakeholders we interviewed were satisfied with many aspects of DOT's selection process, they cited two areas of the TIFIA selection process that they found to be less satisfactory—DOT's application of selection criteria and the uncertainty of the timing of the process. Twenty-seven of the 36 recent TIFIA applicants that responded to our survey indicated that they were satisfied with DOT's explanation of the application process in funding notices, and 28 of the recent applicants reported that the LOI format allowed them to provide sufficient detail about their project.[44] In addition, several applicants told us that the TIFIA selection process was fairly simple to understand and not overly burdensome, and many applicants and advisors we interviewed told us that they found the TIFIA staff to be very cooperative and helpful. Moreover, many recent applicants told us that they appreciated that DOT gave feedback to the sponsors of unsuccessful LOIs.

Recent applicants we surveyed and interviewed, however, expressed two primary concerns about the process: the lack of transparency with which DOT applies statutory criteria to select LOIs to advance, and the uncertainty of the timing of the process. Only 7 of the recent applicants that responded to our survey were satisfied with the transparency of DOT's project selection decisions. In addition, some recent applicants we interviewed and surveyed indicated that clarifications made to two criteria—national or regional significance and environment [45]—are highly subjective, and that it is difficult to understand what characteristics DOT is using to determine the extent to which a project does or does not meet these criteria. For example, one survey respondent indicated that it is unclear how one qualified project is selected over another in the competitive process. In addition, one recent applicant we interviewed said that it does not know what characteristics DOT looks for or uses to determine if a project does or does not meet a criterion, particularly for the livability clarification in national or regional significance. Some recent applicants also indicated that the LOI evaluation and selection process remained unclear, even after receiving feedback from DOT. Of the 21 recent applicants that indicated they received feedback, 8 reported that it was slightly or not at all useful in understanding the scoring of their LOIs—the primary aim of feedback.

Several financial and legal advisors as well as private concessionaires we interviewed also said that there is a lack of transparency in the application of the criteria. These advisors indicated that DOT could be more transparent about the selection criteria and scoring process it uses to select projects. As we reported previously regarding competitively selected funding programs, were DOT to make additional information on its selection decisions publicly available, potential applicants would have better information on how to create and submit well-developed projects.[46] When such information is not made available, DOT may invite speculation that projects were selected for reasons other than merit.

In addition, recent applicants and financial and legal advisors we interviewed said the timing of the LOI evaluation and selection process is inconsistent from year to year and therefore creates uncertainty. Specifically, several applicants and advisors we interviewed told us that the inconsistent timing in both the dates of the release of DOT's funding notice and LOI submission deadline, as well as the announcement of the outcome of the selection process, contributes to this uncertainty. [47] Because TIFIA projects are typically high-cost projects with multiple funding and financing streams, the uncertainty about when a project can submit an LOI and more importantly, when a project can count on a TIFIA credit agreement to fill a funding gap, can affect the financial feasibility of these projects. For example, one financial advisor said that because the current LOI process occurs only once per year, it makes it difficult to plan and to coordinate with other vital project planning pieces, like state budget cycles, environmental reviews, and private investors' timelines. If a project sponsor misses the solicitation for a particular year, it has to wait another year to submit an LOI. The uncertainty as to when the outcome of the selection process will be announced can also affect projects. According to one financial advisor, project delays can affect construction costs or public support for the project, among other things. DOT officials said that the timing of the annual solicitation is due, in part, to receiving budget authority for the TIFIA program on a year-to-year rather than a multiyear basis. On the basis of feedback from fiscal year 2010 applicants, DOT has also tried to shorten the LOI evaluation and selection process in subsequent years so that applicants learn outcomes sooner. For fiscal year 2010, this process, measured from the date LOIs were due to the announcement of which LOIs were invited to apply, took about 6 months, but for fiscal years 2011 and 2012, the process took 5 months and 4 months, respectively.

DOT Is Taking Steps to Improve the LOI Evaluation Process

DOT has made changes to try to improve the LOI evaluation process since returning to a competitive fixed-date selection process. In particular, DOT officials said that they are applying best practices from other DOT discretionary programs such as the TIGER program and learning from past rounds of TIFIA solicitations. The changes include the following:

- In fiscal year 2011, DOT increased its documentation of key decisions for the LOI evaluation and selection process. For each LOI, the multimodal team summarized its deliberations on the extent to which a project met each statutory criterion in a standard form. In addition, to aid in providing feedback to unsuccessful applicants, TIFIA office staff produced an internal memo to document the multimodal team's rationale for LOI scores and grouping as well as the executive leadership team's concurrence with these evaluations.
- In fiscal year 2012, DOT made changes to the LOI evaluation process at the multimodal team level.[48] Specifically, the team assigned qualitative scores—"not aligned," "somewhat aligned," "well aligned," and "very well aligned"—rather than numeric scores to LOIs for each criterion. According to DOT officials, these changes facilitate discussion within the team and accelerate progress to consensus on project scores and impressions. In addition, DOT officials said the qualitative scores are more reflective of the actual evaluation process than the numeric scoring system used in past rounds of solicitation and prevent the team from focusing too heavily on the numeric scores.
- In fiscal year 2012, DOT further clarified the TIFIA funding notice. In particular, DOT included the two statutory selection criteria that had not been considered as part of the LOI selection process for fiscal year 2010 or 2011—creditworthiness and consumption of budget authority. It also stated that in selecting LOIs to advance, it may give priority to projects that enhance the geographic diversity of the TIFIA portfolio and may consider the project's readiness and timeline to proceed to financial close. DOT officials said it did so as part of its efforts to improve its communication of the criteria and selection process to applicants through the funding notices over time.
- For fiscal year 2012, DOT invited the sponsors of five projects to apply for TIFIA credit assistance. In addition, in response to concerns raised by project sponsors as well as the lack of certainty about future funding levels associated with the TIFIA program because of the absence of a long-term surface reauthorization, DOT officials said that an expedited review process would be created for additional highly-rated projects if TIFIA budgetary resources are significantly increased based on the President's Budget Request for fiscal year 2013.[49]

Since many of DOT's changes to the selection process occurred in the fiscal year 2012 TIFIA solicitation, it is too soon to know whether these changes will address the transparency and uncertainty concerns raised by recent applicants and financial and legal advisors. DOT officials said that they will continue to explore other changes to the process, such as creating additional internal guidance on scoring projects or changing feedback. Additionally, DOT officials said that the variety of TIFIA projects by size and mode could make it difficult for DOT to specify how particular benefits translate to a score for an LOI. For instance, in 2010 sponsors submitted LOIs for projects that varied greatly in terms of benefits, size, and mode, as exemplified by the $360 million Southeast Waterfront project—a 5-mile bus, auto, bicycle, and pedestrian corridor that is part of a redevelopment project in San Francisco—and the $1.5 billion Goethals Bridge project—the replacement of a existing bridge connecting New York and New Jersey. DOT officials said that the current LOI selection process was developed in response to the combination of high demand and uncertain budgetary environment, and indicated that it would likely modify the evaluation and selection process in response to an increase in TIFIA's budget authority.

FLEXIBLE REPAYMENT TERMS AND OTHER FACTORS INFLUENCE WHETHER STATES AND OTHER SPONSORS SEEK TIFIA ASSISTANCE; FUTURE DEMAND IS DIFFICULT TO GAUGE

The TIFIA program's flexibility and low interest rates are the predominant reasons why sponsors seek TIFIA assistance. TIFIA's flexibility extends to both repayment terms and debt structuring. For states that have not sought TIFIA assistance, state DOTs indicated that a variety of factors contributed to their decision not to use TIFIA, such as a lack of projects that met the eligibility requirements or the availability of other financing options. Looking ahead, future demand for TIFIA is difficult to gauge because it is influenced by a number of factors such as changes to interest rates or state fiscal conditions.

A Number of Factors Influence whether States and Sponsors Seek TIFIA Assistance

As shown in table 5, most recent applicants we surveyed cited TIFIA's repayment terms and options, low interest rate, and ability to serve as subordinate debt as very or somewhat important in their decision to seek assistance in fiscal years 2010 and 2011.[50]

In addition to recent applicants we surveyed, other project sponsors, financial and legal advisors, and private concessionaires we interviewed consistently cited TIFIA's flexible terms as a major benefit of the program. According to DOT officials, the major benefits of TIFIA are that it can be a patient, flexible lender and can help a sponsor secure a portion of the project's lending to attract other financing.

Table 5. Top Factors That Affected Recent Applicants' Decisions to Seek TIFIA Assistance

	Very or somewhat important	Neither important nor unimportant	Very or somewhat unimportant	Don't know/not applicable/ missing	Total
Repayment terms and options offered by TIFIA credit assistance	34	0	0	2	36
Low interest rate of TIFIA relative to other financing options	34	0	1	1	36
Ability of TIFIA credit assistance to serve as subordinate debt in a project finance plan[a]	32	0	3	1	36

Source: GAO survey of recent TIFIA applicants.

[a] TIFIA assistance can be subordinate to a project's senior debt in a project finance plan, meaning that senior creditors may receive project revenue ahead of DOT, except in the event of a borrower's bankruptcy, insolvency, or liquidation.

For example, one project sponsor said that deferring payment for 5 years after substantial completion is very important for new toll road projects to allow time for usage to grow and thus revenues to ramp up after opening. Beyond favorable repayment terms, TIFIA assistance can be subordinate to other debt, meaning that this other debt may receive project revenue ahead of DOT except in the case of bankruptcy, insolvency, or liquidation.[51] All six of the private concessionaires we interviewed said that this structure is a key benefit of the program, as it can help improve their ability to raise senior debt. Many project sponsors we spoke to also cited TIFIA's relatively low interest rate as a main benefit of the program. The interest rate of TIFIA assistance is based on U.S. Treasury securities of a similar maturity and, since 2008, these Treasury rates have been lower than municipal bond interest rates.

To a lesser extent, recent applicants we surveyed cited several other factors as important in their decision to seek TIFIA assistance in fiscal years 2010 and 2011. (See table 6.)

Table 6. Other Factors That Affected Recent Applicants' Decisions to Seek TIFIA Assistance

	Very or somewhat important	Neither important nor unimportant	Very or somewhat unimportant	Don't know/not applicable/ missing	Total
Availability of user fees to repay TIFIA credit assistance	27	2	4	3	36
Ability to accelerate delivery of project through use of TIFIA credit assistance	27	4	3	2	36
Lack of other funding options	26	5	4	1	36

Source: GAO survey of recent TIFIA applicants.

For instance, survey responses indicate that the TIFIA program can provide financing to projects that is unavailable in the financial markets, particularly for projects with unproven revenue streams. For example, one project sponsor we interviewed that received a TIFIA loan said that obtaining subordinate debt in the financial markets would have been prohibitively expensive, since the project was a new toll road. In addition, the ability of TIFIA to help accelerate the delivery of projects was also important among recent applicants. Officials from the Florida Department of Transportation estimated that its TIFIA loan helped accelerate completion of the Miami Intermodal Center by 10 years.

Two other project sponsors we interviewed said TIFIA plays an important role in accelerating not only the projects for which they received assistance but other major capital projects too, as TIFIA assistance helps free up funds for other projects.

However, our survey also indicated that some states have neither sought nor plan to seek TIFIA assistance. In states where sponsors have never sought TIFIA assistance, the extent to which certain factors affected state DOTs' decisions to not seek credit assistance varied, but many of these state DOTs indicated that they have not submitted LOIs because they (1) do not have projects that meet the eligibility requirements including the required cost threshold, (2) get financing from other source(s), or (3) have state restrictions on borrowing funds for transportation projects.52 (See table 7.)

Table 7. Factors That Affected State DOTs' Decisions to Not Seek TIFIA Assistance

	Very great or great impact	Moderate impact	Some or little to no impact	Don't know/not applicable/ missing	Total
A lack of either user fees or dedicated revenues to repay assistance	8	1	6	1	16
No projects in the state met the required cost threshold ($50 million)	6	1	7	2	16
Other financing options were available	8	2	5	1	16
State is subjected to restrictions on borrowing[a]	6	0	9	1	16

Notes: For the remaining factors on the survey, at least 9 of 16 state DOTs indicated that the factor had some or little to no impact on its decision not to submit an LOI. See appendix III for a table of the survey results for this question.

[a]These restrictions include, among others, that states do not have authorizing legislation to allow its governments or public agencies to borrow funds, or that states require legislative approval to borrow funds through bonds and other financing tools.

Regarding lack of eligible projects, one state DOT indicated in the survey that a lack of dedicated revenues or private investment prevents TIFIA from being a viable option for rural states now and in the future. Several financial and legal advisors we interviewed also said that some states lack projects that are large enough to benefit from the TIFIA program. For the TIFIA program, a sponsor must pay a $50,000 application fee if invited to apply after the LOI stage. Then, if selected to receive assistance, the TIFIA borrower must pay a transaction fee, typically between $300,000 and $400,000, to cover the costs incurred by DOT to negotiate and execute the credit agreement, like costs for external advisors. Borrowers can also incur additional costs from hiring their own advisors and obtaining a credit rating for the project. Therefore, the cost of applying for TIFIA may outweigh the benefits of TIFIA for lower-cost projects.

Moreover, DOT officials as well as several financial advisors, a private concessionaire, and an industry association we spoke to said that the TIFIA program may be better suited to states with more urban populations and a greater need for large-scale projects. States with sponsors that have never sought TIFIA assistance tend to have a smaller portion of their population living in urban areas—that is, areas with a total population of 50,000 or more—than states with sponsors that have sought TIFIA assistance.[53] DOT officials said that to date, TIFIA projects have been located in states with large urban areas that have major transportation needs and can more easily charge tolls or generate other project revenues.

Based on our survey, it is unlikely that many of the states that have not sought TIFIA assistance will seek such assistance in the future. Of the 16 state DOTs from states that have never sought assistance that responded to our survey, only 1 indicated that it anticipated seeking TIFIA assistance in the next 5 years. In addition, most of these state DOTs indicated that changes to the program—such as making more funds available for the program or increasing the portion of project costs that TIFIA assistance could cover—would have somewhat or little to no increase on the likelihood that they would seek TIFIA assistance.[54]

Future TIFIA Demand Exists, but Its Magnitude Is Difficult to Gauge

According to our interviews with DOT, project sponsors, advisors, and private concessionaires, overall demand for the TIFIA program is likely to continue. However, the magnitude of this demand is difficult to estimate because it is influenced by a variety of external factors like changes to interest rates, use of public-private partnerships, and state fiscal conditions.

Changes to the TIFIA interest rate relative to municipal debt interest rates could considerably affect the demand for TIFIA credit assistance. For the last 3 fiscal years, sponsors submitted LOIs for credit assistance totaling more than 10 times what the program's current budget authority can support. Several legal and financial advisors we interviewed said that many project sponsors sought TIFIA in recent years because of depressed market conditions and attractive TIFIA interest rates, relative to interest rates on municipal debt, and a few of these advisors and one industry association said that demand for TIFIA will likely decrease if TIFIA interest rates become less attractive relative to municipal debt interest rates. The relatively low TIFIA interest rates made the program attractive to a greater number of sponsors, even those with access to other financing options. For example, one recent applicant we interviewed said that in the past, TIFIA was a more expensive finance option than issuing its own debt, and its interest in TIFIA during the last few years is primarily driven by the program's relatively low interest rates. The applicant noted that should interest rates on TIFIA loans increase in the future, it will likely seek financing in the private capital markets.

Two other factors will influence the demand for TIFIA assistance.

- Greater use of public-private partnerships and other alternative project delivery approaches could result in a greater demand for TIFIA credit assistance. Many private concessionaires we interviewed said that TIFIA is an important financing tool for public-private partnerships. According to DOT officials, TIFIA credit assistance has been part of the financing package for most large-scale public-private partnership projects in the United States in recent years. In addition, some states, like Colorado and Virginia, have set up offices to facilitate public-private partnerships, so sponsors in such states may be more likely to use this approach given this support.
- State-specific conditions will also influence the demand for TIFIA assistance. As federal and state fuel taxes may not be a sustainable long-term source of transportation funding, state DOTs may make greater use of finance tools like TIFIA to deliver projects. We have previously reported that state and local governments face persistent and long-term fiscal pressures.[55] At the same time, estimates to repair, replace, or upgrade aging transportation infrastructure—as well as expand capacity to meet increased demand—top hundreds of billions of dollars. As a result, DOT anticipates more demand for the TIFIA program as states and localities look to leverage limited funds. One

state we interviewed, for example, said that pay-as-you-go funding—a more traditional means of funding transportation infrastructure whereby a sponsor builds projects in phases or increments as funds are available—no longer keeps pace with infrastructure needs. Therefore, the state DOT has turned to TIFIA to help finance big, high-cost projects that need federal assistance to advance. Looking ahead, 15 of the 42 state DOTs that responded to our survey indicated that they have projects for which they will likely seek TIFIA in the next 5 years. Most of these state DOTs (13) have sought TIFIA assistance in the past and indicated that they are likely to seek TIFIA for 1-5 projects, while a few indicated they are likely to seek TIFIA for 6-10 projects.[56]

Options Proposed to Modify the TIFIA Program

With the pending reauthorization of the surface transportation programs, the tight budgetary environment, and the increase in demand for TIFIA, government and industry officials have proffered options to modify the program.

We reviewed surface transportation reauthorization bills—H.R. 7, the American Energy and Infrastructure Jobs Act as reported by the House Committee on Transportation and Infrastructure, and S. 1813, Moving Ahead for Progress in the 21st Century Act (MAP-21) as adopted by the Senate, respectively—to identify proposed changes to the program. Based on our interviews with select project sponsors, financial and legal advisors, and others, as well as our survey of state DOTs and recent applicants, we identified several recurring options that have been proposed to modify the TIFA program.

Some options require congressional action to implement, while others would require DOT to change program-level policies. Each option has advantages and disadvantages, and thus implementing any of these options would require policy trade-offs. Moreover, some options could affect the overall demand for the program and the sphere of projects that could apply for or benefit from TIFIA.

Table 8 provides a list of proposed options to modify the TIFIA program in the surface transportation reauthorization bills and the President's fiscal year 2013 budget.

Table 8. Summary of Proposed Changes to the TIFIA Program

Proposed change	Fiscal year 2013 President's budget	American Energy and Infrastructure Jobs Act of 2012[a]	MAP-21[a]
Increase authorized budget authority	□	□	□
Increase funding for program administration	□	□	□
Explicitly allow any project sponsor to pay subsidy cost		□	
Increase the portion of costs covered by TIFIA (from 33% to 49%)[b]		□	□
Allow exceptions to the nonsubordination clause		□	□
Remove all selection criteria		□	□
Return to a rolling application process		□	□
Allow program of projects to apply[c]		□	□
Decrease the required total project costs required for rural infrastructure projects		□	□
Expand eligible costs (e.g., development phase and preconstruction costs)		□	
Offer lower interest rates to rural infrastructure projects			□
Require two credit ratings agencies to rate senior debt		□	□
Require DOT to notify applicants on the status of their applications within mandated timeframes		□	

Source: GAO analysis of proposals and legislation.

[a] This analysis is based on H.R.7 The American Energy and Infrastructure Jobs Act of 2012 as reported to the House of Representatives by the House Committee on Transportation and Infrastructure (H.R. 7, 112th Cong. (Feb. 13, 2012)) and S. 1813 the Moving Ahead for Progress in the 21st Century Act (MAP-21) as adopted by the Senate (S.1813, 112th Cong. (Feb. 6, 2012)).

[b] In MAP-21, TIFIA lines of credit may not exceed 33 percent of the anticipated eligible project costs.

[c] Currently only individual projects with dedicated revenues may apply for credit assistance through the TIFIA program. This proposed change would allow a set of related projects that will be phased in over a period of several years to apply and be approved under a single master credit agreement. As with individual projects, a program of projects would have to meet other statutory requirements before entering into a master credit agreement.

Increase the Number of Projects the TIFIA Program Can Support

Two proposed changes, increasing the amount of authorized budget authority and allowing project sponsors to pay fees to contribute to the credit subsidy cost, could potentially allow the TIFIA program to provide more assistance to projects.

Increase amount of authorized budget authority. Members of Congress, DOT, and others have proposed increasing the amount of authorized budget authority to cover the subsidy costs for the TIFIA program. Proposals vary from increasing this amount to $1 billion, as in the reauthorization bills, to a smaller increase of $500 million proposed in DOT's fiscal year 2013 budget. Congressional support for an increase in authorized budget authority for TIFIA is rooted in the program's ability to leverage funds and stretch federal dollars further than a traditional grant program. These proposals represent significant increases to TIFIA's current annual authorized budget authority of $122 million. Proposals to increase the amount of authorized budget authority for the TIFIA program occur during an austere federal budget environment. The Budget Control Act of 2011 places limits on discretionary spending for the next 10 fiscal years.[57] As a result, an increase in one area of discretionary spending, like the TIFIA program, requires a decrease in another area of discretionary spending.

Increasing the amount of authorized budget authority is strongly supported by recent applicants we surveyed as well as legal and financial advisors we interviewed. For example, 32 out of 36 recent applicants that responded to our survey strongly support expanding funding for the TIFIA program. An increase in funding would likely allow the program to provide more credit assistance, in terms of the number of projects receiving credit assistance or the amount of credit assistance provided to each project. An increase in funding could also allow the program to come closer to meeting the current demand for the program, which is more than 10 times what the current budget authority could support. However, DOT officials and other stakeholders told us that an increase in funding would need to be accompanied by an increase in administrative resources. According to project sponsors and other stakeholders, the TIFIA office has been very responsive and helpful, but a few said that response time has slowed in recent years. With increased funding, DOT would likely see an increase in the number of applications to review, credit agreements to negotiate, and credit agreements to monitor. DOT officials said they are prepared to adjust staffing levels in the event that Congress provides the TIFIA program with an increase in authorized budget authority as is proposed in the surface transportation reauthorization bills. Further, DOT officials said that an increase in TIFIA funding may require DOT to reexamine how it manages the program—such as how it selects projects and negotiates credit agreements—and issue new regulations.

Allow sponsors to pay fees to contribute to the credit subsidy cost of assistance. H.R. 7 would mandate that DOT allow project sponsors to pay fees to reduce the credit subsidy cost of assistance if DOT funds run out.[58] According to DOT, current law allows but does not require DOT to let the approved sponsor pay a fee to reduce the credit subsidy cost of the project in the event that there is insufficient budget authority to fund credit assistance for a selected TIFIA project. Over the life of the TIFIA program, three project sponsors have paid fees to reduce the credit subsidy cost of their TIFIA assistance; all three cases occurred after the program became oversubscribed in fiscal year 2008.

Among recent applicants we surveyed and project sponsors we interviewed, many supported this program change.[59] Supporters of this option said that given the high demand for TIFIA credit assistance and limited budget authority, allowing project sponsors to pay fees to cover the credit subsidy cost when DOT's budget authority runs out would allow more eligible projects to be built and reduce the oversubscription of the program. However, DOT previously decided against instituting this option more broadly through a pilot program in 2010. DOT officials told us that while allowing project sponsors to pay fees to cover the credit subsidy cost provides flexibility, especially when demand outpaces budget authority, it complicates the negotiation of credit agreements. While DOT would have to follow its subsidy estimation methodology to determine a project sponsor's fee, the project sponsor may want to negotiate the fee. Project sponsors we interviewed said that for this option to work, DOT would need to provide them with more information on how the credit subsidy cost is calculated. Under FCRA, OMB is responsible for subsidy cost estimates. OMB may delegate this authority to the agency providing credit assistance, but the delegation should be based on the written guidelines or criteria developed by OMB.[60] OMB retains the responsibility and final approval of subsidy cost estimates. Given these complexities, this option may be difficult to implement, though it could be done relatively quickly. In addition, to the extent that DOT underestimates the initial subsidy costs and does not collect enough fees from borrowers, taxpayers will ultimately have to pay for any shortfalls.[61]

Allowing project sponsors to pay fees to cover the credit subsidy cost could remove the congressional limit on the size of the TIFIA program and thus increase the federal government's exposure. According to DOT, SAFETEA-LU removed the cap on the amount of credit assistance the TIFIA program could provide each year, so the only limit on the TIFIA program's size currently is the budget authority provided by Congress. DOT officials said that allowing project sponsors to pay the subsidy cost could allow the program to grow larger than Congress authorized through budget authority. DOT officials told us that for other DOT credit programs, such as the Railroad Rehabilitation and Improvement Financing (RRIF) program, project sponsors are required to pay fees towards the credit subsidies for loans because they do not have budget authority for this purpose, but the RRIF program has a statutory limit on total outstanding credit assistance.[62] When Congress imposes such a limit, it can control the government's exposure to financial losses. Moreover, if this proposed change were adopted in combination with other proposed changes to the program—requiring the Secretary to approve all qualifying applications— the total size and exposure of the TIFIA program could expand dramatically.

Increase the Portion of Costs Covered by TIFIA

Another option in the reauthorization bills would increase the portion of eligible project costs TIFIA assistance could cover from 33 percent to 49 percent.[63] Among think tank and industry group proposals, project sponsors, and other stakeholders we interviewed, support for this option varied. Those that support increasing the TIFIA share said it would reduce the burden on sponsors to find nonfederal sources of debt and allow them to borrow more funds on favorable terms. For example, several project sponsors said that for very large infrastructure projects, finding a combination of federal, state, and private financing can be difficult. Those that do not support this option expressed concern that it would reduce the incentive to find private and other nonfederal financing and potentially reduce market discipline that comes from other lenders to projects. For example, several stakeholders we interviewed said that increasing the percentage of total project costs that TIFIA can finance could result in project sponsors substituting TIFIA credit assistance for private debt or private equity investments. Others expressed concern that increasing the share of costs TIFIA covers would potentially reduce the availability of

TIFIA assistance, especially if Congress does not increase budget authority for the TIFIA program.

Moreover, increasing the portion of costs covered by TIFIA could decrease the program's ability to achieve one of its key goals—leveraging federal funds. DOT officials told us that changing the statute to increase the TIFIA share could reduce the number of projects supported (for a given amount of budget authority) and reduce the leveraging of federal funds as project sponsors seek more financing through TIFIA rather than other sources. Currently, DOT estimates that each $10 million in budget authority can provide up to $100 million in TIFIA credit assistance and leverage $300 million in transportation infrastructure investment. If the limit on TIFIA assistance were increased to 49 percent, this same amount of budget authority could leverage about $200 million in transportation infrastructure investment.[64] This change could also increase the exposure of the federal government to the risk of loan defaults if the size of the credit assistance for each project increases.

Eliminate or Waive the Nonsubordination Clause

The reauthorization bills propose exceptions to the nonsubordination clause. For example, the Senate reauthorization bill, S. 1813, allows exceptions to the nonsubordination clause for certain types of borrowers.[65] Specifically, public agencies that are financing ongoing capital programs and have senior bonds outstanding could be exempt from the nonsubordination clause if (1) the outstanding bonds are rated A or higher, (2) the TIFIA assistance and outstanding bonds are secured by revenues not affected by project performance (e.g., sales tax), and (3) the TIFIA assistance is 33 percent or less of the total project costs. Among recent applicants we surveyed, 22 out of 36 strongly or moderately support allowing waivers to the nonsubordination clause. Several legal and financial advisors and other project sponsors we interviewed support removing the nonsubordination clause altogether.

While there is general support for allowing waivers to or eliminating the nonsubordination clause, many of those we interviewed indicated that the clause does not pose an insurmountable challenge to negotiating a credit agreement, and that it provides needed protection for the federal government. Eliminating or waiving the nonsubordination clause could address some issues identified by financial advisors and credit ratings agencies we interviewed. For example, the TIFIA nonsubordination clause can be difficult to integrate with existing terms for outstanding bonds secured by the same revenue stream. If the nonsubordination clause is triggered due to project bankruptcy or insolvency, project sponsors must make special arrangements to ensure this bond covenant is not violated. However, despite these issues, many project sponsors and legal and financial advisors said that the nonsubordination clause provides an important protection to taxpayers. Moreover, few if any could point to instances where it prevented the closing of a credit agreement.

DOT officials said that the nonsubordination clause helps protect the federal government and taxpayers. For the TIFIA program, the nonsubordination clause is used to lessen the risk to the federal government.[66] While the nonsubordination clause can cause issues for borrowers, DOT officials said that they can work with borrowers to try to address financial difficulties before they must legally invoke the clause. For example, DOT can defer invoking the nonsubordination clause for up to a year after a missed payment, but to date no sponsor has missed a payment. In addition, DOT officials told us that removing the nonsubordination clause would increase the federal government's risk because it would lower the likelihood of recovering funds. According to DOT officials, the nonsubordination clause facilitated its involvement in bankruptcy discussions for the South Bay Expressway and, as a result, DOT expects to recover, through the restructuring of the project's debt and assumption of the loan by SANDAG, up to 100 percent of the original loan value. Further, DOT officials said that without the nonsubordination clause, the credit subsidy cost required for a project would increase significantly, because of the increased risk to the federal government, and thus reduce the amount of assistance the TIFIA program could provide.

Modify the Selection Process

Modify selection criteria. Both the reauthorization bills propose eliminating TIFIA's selection criteria and adding to the current eligibility requirements.[67] H.R. 7 would expand eligibility requirements to include creditworthiness, regional significance, beneficial effects, and project readiness.[68] S. 1813 adds creditworthiness to the program's current eligibility requirements.

Twenty-three of 36 recent applicants that responded to our survey support modifying the TIFIA selection criteria, but when asked how the criteria should be modified, these respondents most often indicated that they want more transparency in how selection criteria are applied.[69]

Project sponsors and advisors we interviewed said they would prefer more transparency in the evaluation of LOIs and a better explanation of how selection criteria are applied. Several project sponsors and advisors expressed concern about the definitions of some criteria—in particular, the livability and sustainability clarifications—as well as how the criteria are applied to LOIs.

However, altering or eliminating the selection criteria could modify the nature of the TIFIA program, changing it from a discretionary program where select projects receive assistance to more of an eligibility-based program where all eligible, creditworthy projects can receive assistance. DOT supports retaining the statutory criteria to use in selecting projects to receive credit assistance. In this way, the TIFIA program would continue to provide assistance to projects that meet DOT's national transportation goals. DOT officials added that just because a project is creditworthy does not ensure that it will have positive transportation benefits. Further, modifying or eliminating the selection criteria could be implemented in several ways, each entailing different trade-offs. For example, one of the new eligibility requirements in H.R. 7 is "beneficial effects," which collapses some existing statutory criteria and program goals—specifically, fostering public-private partnerships, attracting private debt or equity investment, enabling a project to proceed faster than without the credit assistance, and reducing federal grant assistance—into one category. While fostering public-private partnerships, for example, is one of the selection criteria for the current TIFIA program, projects without a public-private component are still eligible to apply. Depending on how the beneficial effects eligibility requirement, if enacted, is defined and implemented, it could render some projects—including some that recently received credit assistance—ineligible.

Return to an open application cycle. The reauthorization bills propose returning to an open application process and prohibiting a fixed-date solicitation. Several project sponsors as well as financial and legal advisors we interviewed support a return to an open application cycle. Some project sponsors said this would allow sponsors to seek TIFIA credit assistance according to a project's schedule, rather than trying to alter this schedule to fit the annual TIFIA solicitation. One state DOT said that the projects applying for TIFIA credit assistance are very complex and must manage multiple timelines for various financing stakeholders, which is further complicated by TIFIA's once-a-year solicitation. A few financial advisors and one project sponsor we spoke with also indicated that increasing the number of solicitations per year would be an improvement if DOT did not return to an open application cycle.

Moreover, due to the fixed-date solicitation process, some project sponsors may be submitting LOIs for projects not yet ready to use TIFIA assistance. DOT previously reported that based on its use of an annual, fixed-date application process from 1999 to 2001, project sponsors may have been applying for assistance prematurely in response to the limited application window. DOT switched to an open application process to allow sponsors to apply based on a project's schedule. For example, one recent applicant told us it submitted an LOI early, as it planned and obtained permits for the project, to familiarize DOT with the project and improve its chances of obtaining TIFIA credit assistance in the next few years. DOT officials told us the fixed-date application cycle is currently a necessity because of limited resources; however, if they had more funds to pay the credit subsidy costs for credit assistance, they would prefer to use an open application system that allows a sponsor to seek TIFIA when it best fits a project's schedule.

Importantly, returning to an open application cycle removes the competitive nature of the TIFIA program. If the TIFIA program's authorized budget authority remains at current levels or does not meet total demand, a project's order in line would determine whether it receives assistance, not its relative merit. Further, if this option were adopted, DOT would have to reconsider its current two-step selection process and determine the extent to which it has the discretion to distribute assistance based on geographic location, project readiness, or other factors not included in the statutory eligibility requirements.

CONCLUSION

Until recently, the innovative credit assistance offered by the TIFIA program to finance the construction of large-scale surface transportation projects was underutilized. However, demand for the program surged, in part because of the tightening of commercial credit markets and low federal treasury interest rates. TIFIA is increasingly becoming a more recognized approach for filling funding and financing gaps for complex transportation projects that can help to mitigate mobility and other transportation issues in many congested urban areas in the United States. DOT, project sponsors, legal and financial advisors, and other stakeholders in the transportation industry have expressed strong support for the program, and Members of Congress have recently developed several reauthorization proposals aimed at greatly increasing the authorized budget authority for the program and modifying other aspects of the program to make it more accessible. DOT has taken some steps to monitor and assess the program through its project oversight and credit monitoring of individual TIFIA credit agreements and, early in the program's tenure, by tracking and reporting on the private investment and leveraging effect of TIFIA to gauge its progress in meeting program goals. However, since that time, DOT has not publicly reported on these or other measures to assess the program as a whole. Without other measures in place going forward, Congress will not have the complete and aggregated data needed to make informed decisions about the program's size and structure.

Additionally, in response to increased demand for the program and multiple extensions of the surface transportation reauthorization over the last 3 years, DOT has had to adapt its process for selecting projects, focusing its review of projects on applicants' LOIs and selecting projects based on their relative merits. The new process, whereby DOT balances a limited program budget authority with selecting projects that are most consistent with the statutory selection criteria, is a work in progress. In response to feedback from applicants and lessons from this and other discretionary programs, DOT has taken steps to make the TIFIA selection process transparent by publicizing the selection criteria and other factors that contribute to project selection and providing feedback to unsuccessful applicants, and many think these steps have been useful. However, many recent applicants and financial and legal advisors that assist applicants in developing projects still feel that the process lacks transparency, making it difficult for them to advance well-developed LOIs. While federal agencies rarely publicly disclose the reasons for their selection decisions in a competitive review process, the considerable demand for TIFIA and changes to the selection process suggest that publicly disclosing additional information about how selection decisions are made would better enable potential applicants to identify how DOT is using the statutory criteria to select projects and develop effective LOIs.

RECOMMENDATIONS FOR EXECUTIVE ACTION

To improve the implementation of the TIFIA program and enable Congress and DOT to better assess program performance, we recommend that the Secretary of Transportation further develop and define performance measures to monitor and evaluate progress toward meeting the program's goals and objectives.

To ensure that future project selections in the TIFIA program are transparent to Congress, applicants, and the public, we recommend that the Secretary of Transportation better disclose information, through notices of funding availability or other program guidance, regarding how DOT evaluates and selects projects.

Agency Comments

We provided a draft of this report to DOT for review and comment. In response, DOT said it would carefully consider the results of our review but did not take a position on whether it agreed with our recommendations. DOT told us that it objectively evaluates applications for TIFIA participation using comprehensive, data-driven processes to identify the most highly qualified projects, and that DOT encourages strong communication with applicants and offers transparent discussion of applicants' submittals to ensure they are fully informed of the basis for program participation decisions. Further, the agency stated that it is continuously reevaluating its processes to ensure they are as effective as possible. The agency also provided technical comments, which we incorporated, as appropriate.

Susan Fleming
Director, Physical Infrastructure Issues

Appendix I: Scope and Methodology

To address our objectives, we reviewed Department of Transportation (DOT) program guidance for the Transportation Infrastructure Finance and Innovation Act (TIFIA) program, relevant legislation and regulations, and DOT's biennial reports to Congress on the TIFIA program. To describe the characteristics and results of the TIFIA program, we analyzed DOT data on past letters of interest (LOI) and applications for assistance to describe the projects that sought credit assistance. We also analyzed data on the projects receiving TIFIA credit agreements through April 2012 to describe these projects, including mode of transport, total cost, amount of TIFIA assistance, amount of private investment, and geographic location. For mode of transport, we used DOT's available characterizations for all projects with credit agreements and for LOIs for fiscal year 2011, while for the remaining LOIs we determined the mode for projects by applying DOT's characterization scheme. When considering the amount of private investment, we followed DOT's convention established in its 2002 report to Congress on TIFIA. Namely, (1) the project must feature investor-held debt or equity and (2) the investment return must be derived from project-generated revenues or other revenues levied specifically to support the project. We only included active credit agreements—those for which sponsors had not repaid or refinanced their credit agreements—as we did not have complete information on the funding sources for all the retired credit agreements. We assessed the reliability of the data by reviewing DOT's data documentation, interviewing knowledgeable officials, and conducting independent validation through use of our web survey. We found the data to be sufficiently reliable for our purposes. In addition, we interviewed DOT officials to learn about the program's goals and the tools DOT uses or plans to use to track and evaluate the performance of credit agreements and the program.

To describe and assess DOT's process for evaluating and selecting projects to invite to apply in fiscal years 2010 and 2011, we examined legislation, regulations, and agency guidance, including notices of funding availability, to describe the statutory and regulatory criteria DOT uses to select projects for credit assistance. We also analyzed and summarized data and documents provided by DOT—including scores assigned and reviewers' assessments of project letters of interest—and interviewed DOT officials to describe the decision-making processes used by the agency to select projects for credit assistance. We focused on federal fiscal years 2010 and 2011, the years for which DOT used a fixed-date competitive solicitation for projects after demand for credit assistance exceeded the program's budget authority and for which the evaluation and selection processes were complete. To assess DOT's process for selecting projects, we compared DOT's process with statute, regulations, and guidance; GAO's *Standards for Internal Control in the Federal Government*; and, as appropriate, past GAO work on federal credit assistance and grant programs.

In addition, we gathered and analyzed data on state-level characteristics, such as federal highway apportionments and whether states have legislative restrictions on borrowing, to determine whether such characteristics were correlated to past demand for the TIFIA program. To explore the potential future demand for TIFIA credit assistance, we analyzed data from DOT on interest in the program in the last 2 fiscal years. To identify the options proposed to modify the TIFIA program, we reviewed reauthorization proposals for surface transportation programs from congressional committees, DOT, and industry and research organizations.

We also interviewed a variety of stakeholders to inform our objectives. We interviewed select current and potential project sponsors (such as state DOTs and transit agencies) to learn about their experiences with the TIFIA selection process and the factors that influenced whether they sought TIFIA assistance. In particular, we interviewed project sponsors in the 5 states—California, Colorado, Florida, Virginia, and Texas—that constitute the majority of TIFIA awards to date, as well as a few states that have had little or no experience with the program—North Carolina and Iowa—that varied in terms of geographic location and legislative authority to borrow and use public-private partnerships. In each of these states, we interviewed the state DOT and all project sponsors that received TIFIA credit assistance as of April 1, 2012.

In addition, we interviewed legal and financial advisors that help sponsors apply for TIFIA credit assistance and private concessionaires that invest in large infrastructure projects to learn about their experiences with the TIFIA program, including the selection process. We also interviewed credit rating agencies and industry associations such as the American Association of State Highway and Transportation Officials (AASHTO) and the American Road and Transportation Builders Association (ARTBA), to learn about their experiences with the TIFIA program and to gain additional information about the types of projects that have sought or received TIFIA credit assistance. In our interviews, we also asked about the factors that would influence future demand for the program as well as options to modify the program and the potential trade-offs of implementing such changes to the TIFIA program. Table 9 lists the organizations we interviewed.

In order to gather opinions of the TIFIA program from the users' standpoint, we designed and administered a web-based survey.

The survey was administered to the state DOTs in all 50 states, the District of Columbia, and Puerto Rico, as well as to all recent applicants that submitted an LOI to the TIFIA program in fiscal years 2010 and 2011.[1] The survey population consisted of four unique groups of respondents: state DOTs from states from which no sponsor had ever applied to the TIFIA program; state DOTs from states from which a sponsor had applied to the TIFIA program but not in recent years—that is, 2010 and 2011; state DOTs who had recently applied to the TIFIA program; and other, non-state DOT organizations who had recently applied to the TIFIA program.

Table 9. List of Interviewees—Sponsors, Financial and Legal Advisors, All Others

AASHTO
ARTBA
Nossaman LLP
White and Case LLP
Jeffrey Parker and Associates
KPMG
Mercator Advisors
Public Financial Management
Cintra
Macquarie
Meridiam
Transurban
Miami Access Tunnel LLC
ACS Infrastructure
Fitch Ratings
Moody's
Standard and Poor's
California DOT
Central Texas Regional Mobility Authority
Colorado DOT
Denver Union Station
Florida DOT
Iowa DOT
Los Angeles County Metropolitan Transportation Authority
North Carolina DOT
North Texas Tollway Authority
Denver Regional Transportation District
Riverside County Transportation Commission
San Diego Association of Governments
Southern California Association of Governments
Texas DOT
Transbay Joint Powers Authority
Virginia DOT

Source: GAO.

Survey respondents were presented with different questions in the survey depending on their past experience with the TIFIA program, and whether or not they were from a state DOT. In general, the survey topics included the following:

- factors contributing to organizations' decision to seek, or not to seek, TIFIA assistance;
- satisfaction with the process for submitting an LOI to the TIFIA program;
- opinions on proposed modifications to the TIFIA program;
- potential future demand for the TIFIA program; and
- characteristics of the state DOTs.

In developing the survey, we took steps to ensure the accuracy and reliability of responses. We cognitively tested the survey with representatives from 5 state DOTs and one other organization included in the respondent population to ensure that questions were clear, comprehensive, and unbiased, and to minimize the burden the survey placed on respondents. On the basis of feedback from the six pretests we conducted, we made changes to the content and format of some survey questions. We obtained contact information for the survey recipients from two sources. First, we obtained contact information for the state DOTs from AASHTO, specifically, from its Standing Committee on Finance and Administration. Second, we obtained contact information for recent applicants from DOT. We also contacted all of the survey recipients in advance, by e-mail, to ensure that we had identified the correct respondents and to request their completion of the questionnaire.

The survey was administered between January 25, 2012, and April 4, 2012. We distributed a link for the survey to the 83 organizations by email and also subsequently e-mailed and telephoned nonrespondents to encourage a higher response rate. Because this was not a sample survey, there are no sampling errors. However, the practical difficulties of conducting any survey may introduce errors, commonly referred to as nonsampling errors. For example, difficulties in how a particular question is interpreted, in the sources of information that are available to respondents, or in how the data were analyzed can introduce unwanted variability into the survey results. We took steps in the development of the questionnaire, the data collection, and the data analysis to minimize these nonsampling errors. Most of the survey questions included close-ended response categories; however, a few survey questions asked respondents to provide a written response to an open-ended question.

When analyzing written responses, one analyst read the responses and assigned them to different categories, while a second analyst reviewed this categorization. We received completed surveys from 66 respondents for an overall response rate of 80 percent. The survey response rates for the four groups of respondents are presented in table 10 below:

We conducted this performance audit from July 2011 to June 2012 in accordance with generally accepted government auditing standards. Those standards require that we plan and perform the audit to obtain sufficient, appropriate evidence to provide a reasonable basis for our findings and conclusions based on our audit objectives. We believe that the evidence obtained provides a reasonable basis for our findings and conclusions based on our audit objectives.

Table 10. List of Survey Recipients and Respondents, by Type of Organization

Type of recipient	Number of responses	Total of recipients number	Response rate
State departments of transportation	42	52	81%
No TIFIA experience	16	18	89
No recent TIFIA experience	14	19	74
Recent applicant/TIFIA experience	12	15	80
Other organizations with recent TIFIA experience	24	31	77
All respondents	66	83	80

Source: GAO.

Appendix II: Projects Receiving TIFIA Credit Assistance

DOT has awarded 27 TIFIA credit agreements to projects through 26 loans and one loan guarantee. Table 11 provides information on each credit agreement, including the name and location of the project receiving assistance, the amount of credit assistance, and the status of the credit agreement.

Table 11. Projects with TIFIA Credit Agreements

Dollars in millions						
Active credit agreements						
Project (location)	*Mode*	*Total project cost*	*Type ofassistance*	*Amount of assistance*	*Primary revenue pledge*	*Status of assistance*
Miami Intermodal Center (Miami, FL)[a]	Intermodal	$1,664	Direct loan	$270	User charges[b]	Drawing funds
Central Texas Turnpike System (Austin, TX)	Highway	3,278	Direct loan	900	User charges	Repaying partial interest
South Bay Expressway-formerly SR 125 South (San Diego County, CA)[c]	Highway	658	Direct loan	140	User charges	Repaying principal and interest
183-A Turnpike (Austin, TX)	Highway	305	Direct loan	66	User charges	Repaying principal and interest
LA 1 Improvements (Leeville, LA)	Highway	372	Direct loan	66	User charges	Deferring interest
Interlink-formerly Warwick Intermodal Station (Warwick, RI)	Intermodal	267	Direct loan	42	User charges	Repaying interest
Pocahontas Parkway/Richmond Airport Connector (Richmond, VA)	Highway	597	Direct loan	150	User charges	Deferring interest
I-495 Capital Beltway HOT Lanes (Fairfax County, VA)	Highway	1,938	Direct loan	589	User charges	Drawing funds
SH 130 Segments 5-6 (Austin, TX)	Highway	1,328	Direct loan	430	User charges	Drawing funds
Intercounty Connector (Montgomery and Prince George's County, MD)	Highway	2,566	Direct loan	516	User charges	Drawing funds
I-595 Corridor Roadway Improvements (Broward County, FL)	Highway	1,834	Direct loan	603	Availability payments[d]	Drawing funds
Triangle Expressway (Raleigh-Durham, NC)	Highway	1,172	Direct loan	386.662	User charges	Drawing funds
Port of Miami Tunnel (Miami, FL)	Highway	1,073	Direct loan	341.037	Availabi-lity	Drawing funds

Dollars in millions						
Active credit agreements						
Project (location)	*Mode*	*Total project cost*	*Type ofassistance*	*Amount of assistance*	*Primary revenue pledge*	*Status of assistance*
					payments	
North Tarrant Express (Dallas-Fort Worth, TX)	Highway	2,047	Direct loan	650	User charges	Drawing funds
Transbay Transit Center (San Francisco, CA)	Transit	1,189	Direct loan	171	Real estate tax increment	Not yet drawing funds
IH 635 Managed Lanes (Dallas-Fort Worth, TX)	Highway	2,615	Direct loan	850	User charges	Disbursing
Denver Union Station Project (Denver, CO)	Intermodal	519	Direct loan	145.6	Sales tax/real estate tax increment	Drawing funds and repaying partial interest
President George Bush Turnpike Western Extension-SH 161 (Dallas, TX)	Highway	1,268	Direct loan	418.4	User charges	Not yet drawing funds
U.S. 36 Managed Lanes/Bus Rapid Transit Project: Segments 1 and 2 (Denver, CO)	Highway	307	Direct loan	54	User charges	Not yet drawing funds
Eagle Project (Denver, CO)	Transit	2,047	Direct loan	280	Tax revenues	Not yet drawing funds
Downtown Tunnel/Midtown Tunnel/MLK Extension (Hampton Roads, VA	Highway	2,089	Direct loan	422	User charges	Drawing funds
Total		*$29,524*	*$7,490.70*			
Retired credit agreements						
Washington Metro Capital Improvement Program (Washington, DC)	Transit	2,324	Guarantee	600	Interjurisdictional funding agreements	Retired
Tren Urbano (San Juan, PR)	Transit	2,250	Direct loan	300	Tax revenues	Retired

Table 11. (Continued)

Dollars in millions						
Active credit agreements						
Project (location)	*Mode*	*Total project cost*	*Type ofassistance*	*Amount of assistance*	*Primary revenue pledge*	*Status of assistance*
Cooper River Bridge Replacement (Charleston, SC)	Highway	675	Direct loan	215	Infrastructure bank loan repayments	Retired
Staten Island Ferries and Terminals (New York City, NY)	Transit	482	Direct loan	159.225	Tobacco settlement revenues	Retired
Reno Transportation Rail Access Corridor-ReTRAC (Reno, NV)	Intermodal	280	Direct loan	50.5	Room and sales tax	Retired
Miami Intermodal Center (Miami, FL)	Intermodal	[e]	Direct loan	269.076	Tax revenues	Retired
Total	*$6,011*		*$1,593.80*			
Total all agreements	*$35,535*		*$9,084.50*			

Source: GAO analysis of DOT data as of April 16, 2012.

[a]This loan, originally for $170 million in credit assistance, was closed on April 25, 2005. An additional $100 million was requested and approved, and the loan was amended on August 1, 2007.

[b] User charges include tolls and rental car customer facility charges.

[c] Under the plan of reorganization approved by the bankruptcy court in April 2011, the restructured principal amount of the loan is $93 million; see project fact sheet for details

[d] An availability payment is a payment made by the public sponsor of a project to the private concessionaire for its responsibility to design, construct, operate, or maintain a project. The payment is based on particular project milestones or facility performance standards, irrespective of usage of the asset, such as ridership or toll revenue.

[e] Project cost included in first listing of the Miami Intermodal Center.

APPENDIX III: RESPONSES TO QUESTIONS FROM GAO'S SURVEY ON TIFIA

We distributed a survey to all state departments of transportation,[1] as well as the organizations who have submitted a letter of interest for TIFIA assistance during federal fiscal years 2010-2011, to gain insight into their experience with and opinions regarding the TIFIA program.

In total the survey went to 83 recipients, and we received completed surveys from 66 of 83 recipients for a response rate of 80 percent. Tables 12-28 below show responses to questions from the survey related to the TIFIA program and project finance.

We also provided examples and definitions for certain terms used in the questions, which are reprinted below.

Survey respondents were presented with different questions in the survey depending on their past experience with the TIFIA program, and whether or not they were from a state DOT.

For example, we only asked organizations that submitted an LOI in 2010 or 2011 (recent applicants) about their experience with the TIFIA evaluation and selection process.

For more information about our methodology for designing and distributing the survey, see appendix I.

Responses to Questions for Recent Applicants to the TIFIA program

Table 12. How Important or Unimportant Were Each of the Following Factor(s) in Your Organization's Decision to Seek TIFIA Credit Assistance for Projects in Federal Fiscal Year 2010 or 2011?

	Very important	*Somewhat important*	*Neither Important nor Unim portant*	*Somewhat Unimportant*	*Very Unimportant*	*Don't know/not applicable*	*Total*
Availability of user fees to repay TIFIA credit assistance	20	7	2	-	4	3	36
Availability of other dedicated revenue stream(s) to repay TIFIA credit assistance	12	7	5	-	7	5	36
Use of TIFIA to attract private equity	12	3	5	1	10	5	36
Low interest rate of TIFIA credit assistance relative to other financing options	28	6	-1		-1		36
Ability of TIFIA credit assistance to serve as subordinate debt in project finance plan	26	6	-1		2	1	36
Repayment terms and options offered by TIFIA credit assistance	30	4	-	-	-	2	36
Political support given to use of alternative project delivery and financing tools	10	13	9	2	2	-	36
Prior experience using alternative	4	10	13	-	6	3	36

	Very important	*Somewhat important*	*Neither Important nor Unim portant*	*Somewhat Unimportant*	*Very Unimportant*	*Don't know/not applicable*	*Total*
financing and delivery tools to implement projects							
Lack of other funding options	13	13	5	2	2	1	36
Ability to accelerate delivery of project through use of TIFIA credit assistance	12	15	4	1	2	1	35

Source: GAO.

Note: We received completed surveys from 36 of 46 recent applicants (78 percent). Totals may not add to 36 because respondents did not all answer every question.

Table 13. How Satisfied or Dissatisfied Were You with Each of the Following Aspects of the Process for Submitting a Letter of Interest (LOI) for the TIFIA Program in Federal Fiscal Year 2010 or 2011?

	Very satisfied	Somewhat satisfied	Neither satisfied nor dissatisfied	Somewhat dissatisfied	Very dissatisfied	Don't know/not applicable	Total
Explanation of the application process in DOT's Notice of Funding Availability (NOFA)	15	12	4	3	1	1	36
Clarity of DOT's selection criteria in the NOFA	6	15	3	9	2	1	36
Ability to provide sufficient project details in LOI format	12	16	5	1	-	1	35

Table 13. (Continued)

	Very satisfied	Somewhat satisfied	Neither satisfied nor dissatisfied	Somewhat dissatisfied	Very dissatisfied	Don't know/not applicable	Total
Length of time it took from submission of LOI to announcement of projects invited to apply	3	15	4	7	5	2	36
Transparency of DOT's project selection decisions	4	3	13	5	8	3	36
Technical assistance on LOI provided by TIFIA office	8	8	10	2	1	7	36

Source: GAO.

Note: We received completed surveys from 36 of 46 recent applicants (78 percent). Totals may not add to 36 because respondents did not all answer every question.

Table 14. Thinking about the Amount of Effort Your Organization Spent, How Much of a Burden, if at All, Were the Following Aspects of Submitting an LOI?

	Great burden	Moderate burden	Some burden	Little to no burden	Don't know/notApplicable	Total
Cost associated with submitting an LOI	-	15	11	9	1	36
Time associated with submitting an LOI	4	15	12	4	1	36

Source: GAO.

Note: We received completed surveys from 36 of 46 recent applicants (78 percent).

Table 15. For Federal Fiscal Year 2010 or 2011, Did You Receive Any Feedback from DOT TIFIA Staff on Any Letters of Interest for TIFIA Credit Assistance?

	Frequency
Yes	21
No	10
Don't know	5
Total	36

Source: GAO.

Note: We received completed surveys from 36 of 46 recent applicants (78 percent).

Table 16. If You Received Feedback on Any of the Following Topics, How Useful, if at All, Was the Feedback You Received from DOT TIFIA Staff?

	Highly useful	Moderately useful	Slightly useful	Not at all useful	Don't know/not applicable	Total
Clarification of selection criteria	4	5	4	4	4	21
Scoring of your LOI	3	6	6	2	4	21
Areas for improvement on future submissions	6	6	1	2	6	21

Source: GAO.

Table 17. How Much Do You Support or Oppose the Following Proposed Changes to the TIFIA Program?

	Strongly support	*Moderately support*	*Neither support nor oppose*	*Moderately oppose*	*Strongly oppose*	*Don't know/not applicable*	*Total*
Expand funding for the TIFIA program	32	-	-	1	2	-	*35*
Change eligibility requirements to decrease total project cost to below $50 million	3	5	9	9	9	1	*36*
Change eligibility requirements to decrease total ITS project cost to below $15 million	-	7	15	3	9	2	*36*
Increase the portion of costs that TIFIA can cover from 33 percent	19	7	4	2	3	1	*36*
Allow waivers for the nonsubordination clause	13	9	8	1	1	3	*35*
Eliminate the nonsubordination clause	9	6	12	1	3	4	*35*
Allow project sponsors to contribute to the subsidy costs of assistance	10	9	10	-	1	5	*35*
Add separate eligibility requirements and terms to encourage rural infrastructure projects to seek credit assistance	3	7	11	4	7	3	*35*

Source: GAO.

Note: We received completed surveys from 36 of 46 recent applicants (78 percent). Totals may not add to 36 because respondents did not all answer every question.

Table 18. Do You Support Modifying the TIFIA Selection Criteria?

	Frequency
Yes	23
No	2
Don't know	10
Total	35

Source: GAO.

Note: We received completed surveys from 36 of 46 recent applicants (78 percent). Totals may not add to 36 because respondents did not all answer every question.

Responses to Questions for State DOTs from States That Were Not Recent Applicants

Table 19. How Much, if at All, Did Each of the Following Factors Impact Your State DOT's Decision to Not Submit an LOI to the TIFIA Program?

	Very great impact	*Great impact*	*Moderate impact*	*Some impact*	*Little to no impact*	*Don't know/not applicable*	*Total*
No projects within the state that met the required cost thresholds for the TIFIA program	4	2	1	2	5	2	*16*
Lack of either user fees or dedicated revenues to repay assistance	6	2	1	3	3	1	*16*
State has no legal authority to use alternative finance and delivery approaches	3	1	1	2	8	1	*16*
State is subjected to restrictions on borrowing	4	2	-	2	7	-	*15*

Table 19. (Continued)

	Very great impact	*Great impact*	*Moderate impact*	*Some impact*	*Little to no impact*	*Don't know/not applicable*	*Total*
Other financing options were available	5	3	2	1	4	1	*16*
State does not use debt financing for transportation projects	2	1	1	1	9	2	*16*
Inclusion of nonsubordination clause in TIFIA credit agreements	1	-	-	-	9	6	*16*
Cost associated with federalizing projects	1	-	1	-	12	2	*16*
Unfamiliarity with the TIFIA program	-	-	2	4	9	1	*16*
Resources needed to apply for and negotiate a TIFIA credit agreement	1	1	2	2	9	1	*16*
Resources needed to maintain an executed TIFIA credit agreement	1	1	1	3	8	1	*15*

Source: GAO.

Note: We received completed surveys from 16 of 18 state DOTs from states from which no sponsor had ever submitted a letter of interest (89 percent). Totals may not add to 16 because respondents did not all answer every question.

Table 20. In Your Opinion, How Much, if at All, Does Each of the Following Proposed Changes Increase the Likelihood That Your Organization Would Seek TIFIA Credit Assistance in the Future?

	Significantly increase	*Greatly increase*	*Moderately increase*	*Somewhat increase*	*Little to no increase*	*Don't know/not applicable*	*Total*
Expand funding for the TIFIA program	2	2	3	5	17	1	*30*

	Significantly increase	*Greatly increase*	*Moderately increase*	*Somewhat increase*	*Little to no increase*	*Don't know/not applicable*	*Total*
Change eligibility requirements to decrease total project cost to below $50 million	2	3	5	3	16	1	*30*
Change eligibility requirements to decrease total ITS project cost to below $15 million	2	2	3	6	16	1	*30*
Increase the portion of costs that TIFIA can cover from 33 percent	4	4	4	5	12	1	*30*
Allow waivers for the nonsubordination clause	2	2	2	4	12	8	*30*
Eliminate the nonsubordination clause	3	2	-	3	14	8	*30*
Allow project sponsors to contribute to the subsidy costs of assistance	2	2	4	6	11	5	*30*
Add separate eligibility requirements and terms to encourage rural infrastructure projects to seek credit assistance	2	4	5	7	9	3	*30*

Source: GAO.

Note: Of the 37 state DOTs from states from which no sponsor had either not submitted a letter of interest during fiscal years 2010 and 2011 or not ever submitted a letter of interest, we received completed surveys from 30 state DOTs (81 percent).

Table 21. In Your Opinion, How Much, if at All, Does Each of the Following Proposed Changes Increase the Likelihood That Your Organization Would Seek TIFIA Credit Assistance in the Future?

(Only with Responses from State DOTs from Which No Sponsor Had Ever Sought Credit Assistance.)

	Significantly increase	*Greatly increase*	*Moderately increase*	*Somewhat increase*	*Little to no increase*	*Don't know/not applicable*	***Total***
Expand funding for the TIFIA program	1	-	1	3	10	1	*16*
Change eligibility requirements to decrease total project cost to below $50 million	2	1	2	2	8	1	*16*
Change eligibility requirements to decrease total ITS project cost to below $15 million	2	1	1	2	9	1	*16*
Increase the portion of costs that TIFIA can cover from 33 percent	3	1	2	4	5	1	*16*
Allow waivers for the nonsubordination clause	2	-	-	3	6	5	*16*
Eliminate the nonsubordination clause	2	-	-	2	7	5	*16*
Allow project sponsors to contribute to the subsidy costs of assistance	2	-	2	3	6	3	*16*
Add separate eligibility requirements and terms to encourage rural infrastructure projects to seek credit assistance	2	3	2	4	3	2	*16*

Source: GAO.

Note: We received completed surveys from 16 of 18 state DOTs from states from which no sponsor had ever submitted a letter of interest (89 percent).

Table 22. Do You Support Modifying the TIFIA Selection Criteria?

Responses	
Yes	11
No	3
Don't know	15
Total	29

Source: GAO.

Note: Of the 37 state DOTs from states from which no sponsor had either not submitted a letter of interest during fiscal years 2010 and 2011 or not ever submitted a letter of interest, we received completed surveys from 30 state DOTs (81 percent). The totals may not add to 30 because respondents did not all answer every question.

Responses to Questions for All State DOTs

Table 23. In the Next 5 years, Approximately How Many Projects Do You Anticipate Will Be Undertaken in Your State That Have a Total Cost of ...

	None	*1 to 2*	*3 to 5*	*6 to 10*	*More than 10*	*Total*
Less than $10 million?	-	-	-	-	39	*39*
$10 to less than $25 million?	-	-	1	5	33	*39*
$25 to less than $50 million?	-	3	6	8	21	*38*
$50 to less than $100 million?	6	8	5	7	12	*38*

Table 23. (Continued)

	None	*1 to 2*	*3 to 5*	*6 to 10*	*More than 10*	*Total*
$100 to less than $150 million?	13	10	5	7	5	*40*
$150 to less than $200 million?	16	7	7	7	2	*39*
$200 million or more?	17	8	8	4	2	*39*

Source: GAO.

Note: We received completed surveys from 42 of 52 state DOTs (81 percent). Totals may not add to 42 because respondents did not all answer every question.

Table 24. Of the Projects Identified in Table 23, for About How Many Do You Anticipate That Your Organization Will Seek TIFIA Credit Assistance?

	Recent TIFIA experience	No recent TIFIA experience	No TIFIA experience	Total
None	2	7	12	21
1 to 2	3	2	1	6
3 to 5	4	3	-	7
6 to 10	2	-	-	2
Don't know	1	1	3	5
Total	12	13	16	41

Source: GAO.

Note: We received completed surveys from 42 of 52 state DOTs (81 percent). Totals may not add to 42 because respondents did not all answer every question.

Table 25. Does Your State DOT Have the Legislative Authority to Use Any of the Following Tools for Transportation Projects?

		Yes	No	Don't know	Total
Design-build	Legislative authority?	35	6	1	42
	If yes, has this authority been used?	32	3	-	35
Public-private partnerships	Legislative authority?	28	11	2	41
	If yes, has this authority been used?	17	11	-	28
Levy tolls for roads, bridges and tunnels	Legislative authority?	24	17	-	41
	If yes, has this authority been used?	16	8	-	24

Source: GAO.

Note: We received completed surveys from 42 of 52 state DOTs (81 percent). Totals may not add to 42 because respondents did not all answer every question.

Table 26. Does Your State Have Access to Any of the Following Revenue Sources, at Either the State or Local Level, That Could Be Used to Repay a TIFIA Loan?

	Yes	No	Don't know	Total
Tolling	20	19	3	42
Sales tax	11	24	5	40

Source: GAO.

Note: We received completed surveys from 42 of 52 state DOTs (81 percent). Totals may not add to 42 because respondents did not all answer every question.

Table 27. Does Your State DOT Have Either Bonding Authority OR Have Access to the Bond Market through Another State Entity?

	Frequency
Response	
Yes	41
Don't know	1
Total	42

Source: GAO.

Note: We received completed surveys from 42 of 52 state DOTs (81 percent). Totals may not add to 42 because respondents did not all answer every question.

Table 28. Which, if Any, of the Following Legislative Restrictions on Conducting Bond Issuances for Capital Projects Apply?

	Yes	No	Don't know	Total
Limitations on the amount that can be issued	31	8	2	41
Limitations on the frequency with which bonds can be issued	7	31	3	41
Bond issuances require legislative approval	23	15	2	40

Source: GAO.

Note: We received completed surveys from 42 of 52 state DOTs (81 percent). Totals may not add to 42 because respondents did not all answer every question.

Glossary/Examples

ITS stands for intelligent transportation system.

The **nonsubordination clause** (also known in the context of the TIFIA program as the springing lien) means that the TIFIA lien on project revenues can be subordinated to those of senior lenders except in the event of bankruptcy, insolvency, or liquidation of the obligor. In such an instance, the TIFIA lien would rise to parity with senior creditors. This provision can be effected through a master trust agreement, an intercreditor agreement, or other agreement entered into at the time of execution of the credit agreement.

Examples of **other dedicated revenue stream(s)** to repay TIFIA credit assistance may include pledged sales taxes, tax increment financing, and availability payments.

The **TIFIA eligibility requirements** are (1) the project shall be consistent with the state transportation plan, if located in a metropolitan area shall be included in that area's metropolitan transportation plan, and shall appear in an approved state transportation improvement program before the DOT and the project sponsor execute a term sheet or credit agreement that results in the obligation of funds; (2) the state, local servicer, or other entity undertaking the project shall submit a project application to the Secretary of Transportation; (3) a project shall have eligible project costs that are reasonably anticipated to equal or exceed the lesser of $50 million or 33 1/3 percent of the amount of federal aid highway funds apportioned for the most recently completed fiscal year to the state in which the project is located (in the case of a project principally involving the installation of intelligent transportation systems (ITS), eligible project costs shall be reasonably anticipated to equal or exceed $15 million); (4) project financing shall be repayable, in whole or in part, from tolls, user fees or other dedicated revenue sources; and (5) in the case of a project that is undertaken by an entity that is not a state or local government or an agency or instrumentality of a state or local government, the project that the entity is undertaking shall be included in the state transportation plan and an approved State Transportation Improvement Program.

The **TIFIA selection criteria** are (1) national or regional significance (including consideration of livability, economic competitiveness, and safety), (2) private participation, (3) environment (including consideration of sustainability and state of good repair), (4) project acceleration, (5) credit worthiness, (6) use of new technology, (7) consumption of budget authority, and (8) reduced federal grant assistance.

Examples of **user fees** to repay TIFIA credit assistance may include tolls and rental car customer facility charges.

End Notes

[1] Pub. L. No. 105-178, 112 Stat. 107 (1998) (codified as 23 U.S.C. ch.6).

[2] Intelligent transportation systems (ITS) projects must have a total cost of at least $15 million.

[3] These two groups were not mutually exclusive, in that some state DOTs are also recent TIFIA applicants. For the purposes of this report, our definition of state DOTs includes those from the 50 states, the District of Columbia, and Puerto Rico.

[4] H.R. 7, 112th Cong. (Feb. 13, 2012).

[5] S. 1813, 112th Cong. (Feb. 6, 2012). 623 U.S.C. §§ 104, 105.

[7] 49 U.S.C. ch. 53.

[8] 23 U.S.C. §§ 603(b)(2), 604(b)(2).

[9] 23 U.S.C. §§ 601(a)(3), 603(a)(4), 604(a)(4). In order to receive TIFIA assistance, the senior debt obligations funding the project (i.e., those obligations having a lien senior to that of the TIFIA credit instrument on the pledged security) must achieve an investment grade rating.

[10] 23 U.S.C. §§ 603(b)(6), 604(b)(8).

[11] Private concessionaires are private sector entities that have entered into agreements with public agencies to allow greater private sector participation in the delivery and financing of transportation projects.

[12] ITS encompasses a broad range of electronics and communication technologies to enhance the capacity and efficiency of surface transportation systems, including traveler information, public transportation, and commercial vehicle operations.

[13] Pub. L. No. 109-59, § 5309(j), 119 Stat. 1144, 1584 (2005).

[14] Budgeting for the cost of credit programs, including the TIFIA program, is governed by the Federal Credit Reform Act (FCRA) of 1990, Pub. L. 93-344, Title V, as amended (codified at 2 U.S.C. §§ 661-661f), which requires federal agencies to receive budget authorities to cover the estimated long-term cost to the government (which includes defaults, delinquencies, and interest subsidies) of providing credit assistance, calculated on a net present value basis and excluding administration costs. The amount of budget authority DOT may use is subject to annual obligation limitations in annual appropriations acts. Of the $122 million in budget authority that has been authorized per year, DOT has the authority to use up to $2.2 million annually to administer the program, and to collect and spend fees, subject to authority being provided in appropriations acts, to cover expenses related to reviewing, negotiating, and servicing credit agreements.

[15] Actual TIFIA lending capacity is subject to the calculation of the estimated subsidy cost for each credit assistance transaction. The amount varies based on the risk profile of the project and the repayment stream. According to DOT, actual original subsidy rates have ranged from less than 1 percent to over 15 percent of the TIFIA credit assistance received.

[16] For fiscal years 1999, 2000, and 2001, DOT used a fixed-date solicitation process for the TIFIA program but used the open, first-come first-served process after these initial years.

[17] Though the program was oversubscribed in fiscal year 2008, DOT did not institute the fixed-date process until fiscal year 2010 because it reserved most of its anticipated fiscal year 2009 and fiscal year 2010 appropriations for existing applicants.

[18] 23 U.S.C. § 602(b)(2)(A). 1947 C.F.R. § 80.15(a).

[20] To date, six projects have been selected to receive TIGER grants to support TIFIA credit assistance. The sponsors for two of these projects have executed credit agreements and the sponsors for the remaining four projects are working with DOT to apply for and negotiate a credit agreement.

[21] To avoid conflicts of interest, DOT prohibits key external personnel (financial and legal advisors) from also representing TIFIA applicants. The TIFIA office is currently conducting a new solicitation for its advisors.

[22] The number of TIFIA credit agreements and projects differs because one project, the Miami Intermodal Center, has three credit agreements for two loans, one of which was amended. See appendix II for a full list of the TIFIA credit agreements and projects.

[23] Other projects, such as construction of visitors' centers and truck technology systems, account for 7 percent of all LOIs.

[24] Though no sponsor from Mississippi has submitted an LOI, the Mississippi DOT submitted an application for TIFIA in September 2008. The Mississippi DOT met the LOI requirement through FHWA's SEP-15 program.

[25] To date, the smallest TIFIA project cost $267 million and received a loan for $42 million (Interlink). The largest TIFIA project cost $3.3 billion and received a loan for $900 million (Central Texas Turnpike System).

[26] The Federal Transit Administration's (FTA) New Starts program, part of the Capital Investment Grant program, is the federal government's primary financial resource for supporting new major transit capital projects that are locally planned, implemented, and operated, such as light rail and bus rapid transit.

[27] During this time, transit projects composed about 18 percent of submitted LOIs, while highway projects composed about 73 percent of submitted LOIs.

[28] In calculating this figure, we used DOT's convention from its 2002 report to Congress on TIFIA. Namely, (1) the project must feature investor-held debt or equity and (2) the investment return must be derived from project-generated revenues or other revenues levied specifically to support the project. We only included active credit agreements— those for which sponsors had not repaid or refinanced their credit agreements—as we did not have complete information on the funding sources for all the retired credit agreements.

[29] An availability payment is a payment made by the public sponsor of a project to the private concessionaire for its responsibility to design, construct, operate, or maintain a project. The payment is based on particular project milestones or facility performance standards, irrespective of usage of the asset, such as ridership or toll revenue. Typically, the private sector concessionaire is the TIFIA borrower when availability payments are pledged to repay assistance.

[30] One project—the Miami Intermodal Center—received two loans. The sponsor pledged different forms of revenue to repay each loan—user fees from rental car users and fuel tax revenues. See appendix II for a full list of the revenues pledged to repay assistance for all projects with TIFIA credit agreements.

[31] All but one of the TIFIA credit agreements are for loans. Each credit agreement includes an annual disbursement schedule and outlines the conditions and forms for drawing funds. For loans, DOT will disburse funds as often as monthly on a reimbursement basis as costs are incurred for eligible project purposes. For the one loan guarantee made to date, no funds were drawn for the project before the credit agreement expired.

[32] 23 U.S.C. § 609.

[33] In the past, we also recommended that a Department of Energy credit assistance program develop performance measures to evaluate program progress. See GAO, Department of

Energy: New Loan Guarantee Program Should Complete Activities Necessary for Effective and Accountable Program Management, GAO-08-750 (Washington, D.C.: July 7, 2008).

[34] GAO, Rail Transit: FTA Programs Are Helping Address Transit Agencies' Safety Challenges, but Improved Performance Goals and Measures Could Better Focus Efforts, GAO-11-199 (Washington, D.C.: Jan. 14, 2011).

[35] GAO, Traffic Congestion: Road Pricing Can Help Reduce Congestion, but Equity Concerns May Grow, GAO-12-119 (Washington, D.C.: Jan. 12, 2012).

[36] The projects selected to receive TIFIA assistance as of 2002 had a total cost of $15.4 billion. The private co-investment in these projects consisted of about $3 billion in debt and $100 million in equity.

[37] DOT officials told us that they also used a competitive, two-step process in fiscal year 2012.

[38] The multimodal team comprised five members representing the Federal Transit Administration, Federal Highway Administration, and the Office of the Secretary of Transportation (OST). This team design meant that LOIs were reviewed by members with subject matter expertise from several transportation modes.

[39] As noted earlier, the eight statutory criteria are National or Regional Significance, Creditworthiness, Private Participation, Project Acceleration, Use of New Technologies, Consumption of Budget Authority, Environment, and Reduced Federal Grant Assistance; however, DOT did not consider the Creditworthiness and Budget Authority criteria when evaluating LOIs in fiscal years 2010 and 2011, since it did not believe project sponsors would have sufficient information regarding these two criteria at the time of the LOI submission. Further, for these two fiscal years, DOT provided additional clarification on two of the criteria through funding notices; specifically, livability, safety, and economic competitiveness for National or Regional Significance and sustainability and state of good repair for Environment.

[40] The executive leadership team is composed of the Deputy Secretary, Under Secretary for Policy, Assistant Secretary for Budget and Program and Chief Financial Officer, and the General Counsel (in 2010 only), all of whom sit on the DOT Credit Council.

[41] In 2010, the executive leadership team invited all seven sponsors with LOIs in category A to give an oral presentation on their projects. In 2011, DOT officials said sponsors were not invited to give presentations because (1) the number of projects in category A increased and (2) the executive leadership team developed familiarity with several projects that had submitted in successive solicitation rounds. According to DOT officials, the executive leadership team relied more heavily on the modal administrations, in particular FHWA division offices and FTA regional offices, to collect additional information on projects in 2011.

[42] In the fiscal year 2012 funding notice DOT further clarified that it would use factors like budget authority and geographic dispersion to select from amongst highly rated projects.

[43] The fiscal year 2010 funding notice stated that the TIFIA program had a limited amount of credit assistance for new applicants that year. According to the Transportation Infrastructure Finance and Innovation Act 2010 Report to Congress, DOT reserved part of its fiscal year 2010 appropriation to cover the 2008 oversubscription to the program. As a result of DOT's invitations, the TIFIA program's fiscal years 2010 and 2011 budget authority is fully committed.

[44] We received completed surveys from 36 of 46 recent applicants (78 percent).

[45] DOT provided these clarifications beginning in fiscal year 2010 in the annual TIFIA funding notices. The livability, economic competitiveness, and safety clarifications are part of the

National or Regional Significance criterion and sustainability and state of good repair clarifications are part of the Environment criterion.

[46] GAO, Surface Transportation: Competitive Grant Programs Could Benefit from Increased Performance Focus and Better Documentation of Key Decisions, GAO-11-234 (Washington, D.C.: Mar. 30, 2011), and Intercity Passenger Rail: Recording Clearer Reasons for Awards Decisions Would Improve Otherwise Good Grantmaking Practices, GAO-11-283 (Washington, D.C.: Mar. 10, 2011).

[47] In fiscal year 2010, the funding notice was released on December 3, 2009, and in fiscal year 2011, the funding notice was released January 19, 2011; for both these fiscal years, LOIs were due in March after an extension. For fiscal year 2012, the funding notice was released on November 3, 2011, with a December 30, 2011, deadline.

[48] In fiscal year 2012, DOT considered all eight statutory criteria in its evaluations. On the basis of its experience in 2010 and 2011, DOT officials believe applicants can provide sufficient information on the creditworthiness and consumption of budget authority criteria that will enable them to be scored in the LOI evaluation process.

[49] The President's Budget Request for fiscal year 2013 and the current reauthorization proposals include significant increases in the amount of authorized budget authority for the TIFIA program.

[50] As noted previously, we received completed surveys from 36 of 46 recent applicants (78 percent). Recent applicants include both state DOTs and other non-state DOT organizations, such as transit agencies, that submitted LOIs to the TIFIA program in fiscal year 2010 or 2011. See appendix III for a complete list of responses to this and other survey questions.

[51] In the event of bankruptcy, insolvency, or liquidation, DOT's claim would be on parity with senior creditors.

[52] We received completed surveys from 16 of 18 state DOTs from states where a sponsor has never submitted an LOI (89 percent).

[53] A metropolitan statistical area (metro area) is a geographic entity defined by the Office of Management and Budget (OMB) for use by federal statistical agencies in collecting, tabulating, and publishing federal statistics. A metro area contains a core urban area of 50,000 or more population, and it includes the counties containing the core urban area, as well as any adjacent counties that have a high degree of social and economic integration (as measured by commuting to work) with the urban core.

[54] For all but the option to add separate eligibility requirements and terms to encourage rural infrastructure projects, half or more of the 16 state DOTs indicated the changes would have somewhat or little to no increase on the likelihood that they would seek TIFIA credit assistance. For this one other option, five state DOTs indicated that the change would significantly or greatly increase the likelihood that that they would seek TIFIA credit assistance in the future. See appendix III for survey results for this and other questions.

[55] GAO, State and Local Governments' Fiscal Outlook April 2011 Update, GAO-11-495SP (Washington, D.C.: Apr. 6, 2011).

[56] Of the remaining 27 state DOTs, 21 state DOTs indicated that they have no projects for which they will likely seek TIFIA in the next 5 years, 5 state DOTs did not know, and 1 did not respond to that item.

[57] Pub. L. No. 112-25, 125 Stat. 240 (2011).

[58] As noted earlier, budget authority is used to pay the credit subsidy cost and administrative expenses of TIFIA credit assistance. The credit subsidy cost, as required by the FCRA, is based on the estimated long-term cost to the government calculated on a net present value basis, excluding administrative costs.

[59] Nineteen of the 36 recent applicants that responded to our survey supported allowing project sponsors to contribute to the subsidy costs of assistance.

[60] 2 U.S.C. § 661b.

[61] Under FCRA, DOT is required to update, or reestimate, the subsidy costs to reflect actual loan performance and changes in expected future loan performance. Shortfalls identified in annual reestimates are funded through a permanent indefinite appropriation and are not subject to the annual appropriations process.

[62] 45 U.S.C. § 822(d).

[63] S. 1813 raises the portion of project costs TIFIA loans and loan guarantees may cover but retains the 33 percent limit for lines of credit. H.R. 7 increases the portion of project costs that may be covered by loans, loan guarantees, and lines of credit to 49 percent.

[64] This estimate assumes that projects receive the maximum amount of credit assistance (either 33 or 49 percent of total project costs) and uses DOT's general rule of thumb that $1 in budget authority can be leveraged to provide $10 in credit assistance.

[65] TIFIA credit assistance is designed with patient repayment terms to encourage private debt as part of the total financing package. To do this, TIFIA assistance can be subordinate to other debt, meaning that this other (senior) debt may receive project revenue ahead of DOT. However, TIFIA credit agreements include, pursuant to the TIFIA statute, a nonsubordination clause (also known as the springing lien), which states that in the event of a borrower's bankruptcy, insolvency, or liquidation, DOT's claims on project revenues will not be subordinate to the claims of other creditors.

[66] OMB Circular A-129, Appendix B, II.3.c. (2) says that to protect the government's interest, the government's claims on assets should not be subordinated to the claim of other lenders in the case of a borrower's default.

[67] The TIFIA program has five statutory eligibility requirements for projects: (1) inclusion in transportation plans and programs; (2) application submitted by state or local government, public authority, or public-private partnership; (3) project costs of equal to or greater than $50 million or $15 million for intelligent transportation system projects; (4) dedicated revenue sources; and (5) public sponsorship of private entities.

[68] In H.R. 7, the proposed regional significance eligibility requirement states that the project shall be regionally significant or otherwise significantly enhance the national transportation system. Currently, one of the statutory selection criteria is national or regional significance, which is the extent to which the project is nationally or regionally significant, in terms of generating economic benefits, supporting international commerce, or otherwise enhancing the national transportation system.

[69] Sixteen respondents provided an open-ended response elaborating on that question, with 6 indicating that they wanted more transparency.

End Note for Appendix I

[1] These two groups were not mutually exclusive, in that some state DOTs are also recent TIFIA applicants. For the purposes of this report, our definition of state DOTs includes those from the 50 states, the District of Columbia, and Puerto Rico.

End Note for Appendix III

[1] For the purposes of this report, our definition of state DOTs includes those from the 50 states, the District of Columbia, and Puerto Rico.

INDEX

A

B

C

D

E

F

G

N

O

P

Q

R

S